# ᴛʜᴇ**new** biology

# Viruses

# ™new biology

# Viruses

The Origin and Evolution of Deadly Pathogens

JOSEPH PANNO, PH.D.

Facts On File
*An Infobase Learning Company*

**VIRUSES: The Origin and Evolution of Deadly Pathogens**

Facts On File, Inc.
An imprint of Infobase Learning
132 West 31st Street
New York NY 10001

**Library of Congress Cataloging-in-Publication Data**
Panno, Joseph.
  Viruses : the origin and evolution of deadly pathogens / Joseph Panno.
    p. cm. — (The new biology)
  Includes bibliographical references and index.
  ISBN 978-0-8160-6855-5
  1. Viruses. 2. Virology. I. Title.    #ANF 2-21-12
  QR360.P36 2011
  579.2—dc22                2010023664

Facts On File books are available at special discounts when purchased in bulk quantities for businesses, associations, institutions, or sales promotions. Please call our Special Sales Department in New York at (212) 967-8800 or (800) 322-8755.

You can find Facts On File on the World Wide Web at http://www.infobaselearning.com

Text design by Erik Lindstrom
Composition by Hermitage Publishing Services
Illustrations by the author
Photo research by Elizabeth H. Oakes
Cover printed by Yurchak Printing, Inc., Landisville, Pa.
Book printed and bound by Yurchak Printing, Inc., Landisville, Pa.
Date printed: April 2011
Printed in the United States of America

10 9 8 7 6 5 4 3 2 1          3 6000 00102 2688

This book is printed on acid-free paper.

# Contents

# Preface

When the first edition of this set was being written, the new biology was just beginning to come into its potential and to experience some of its first failures. Dolly the sheep was alive and well and had just celebrated her fifth birthday. Stem cell researchers, working 12-hour days, were giddy with the prospect of curing every disease known to humankind, but were frustrated by inconsistent results and the limited availability of human embryonic stem cells. Gene therapists, still reeling from the disastrous Gelsinger trial of 1998, were busy trying to figure out what had gone wrong and how to improve the safety of a procedure that many believed would revolutionize medical science. And cancer researchers, while experiencing many successes, hit their own speed bump when a major survey showed only modest improvements in the prognosis for all of the deadliest cancers.

During the 1970s, when the new biology was born, recombinant technology served to reenergize the sagging discipline that biology had become. This same level of excitement reappeared in the 1990s with the emergence of gene therapy, the cloning of Dolly the sheep, and the successful cultivation of stem cells. Recently, great excitement has come with the completion of the human genome project and the genome sequencing of more than 100 animal and plant species. Careful analysis of these genomes has spawned a new branch of biological research known as comparative genomics. The information that scientists can now extract from animal genomes is expected to improve all other branches of biological science. Not to be outdone, stem cell researchers have found a way to produce embryo-like stem cells from ordinary skin cells. This achievement not only marks the end of the great stem cell debate, but it also provides an immensely powerful procedure, known as cellular dedifferentiation, for studying and manipulating the very essence of a cell. This procedure will become a crucial weapon in the fight against cancer and many other diseases.

The new biology, like our expanding universe, has been growing and spreading at an astonishing rate. The amount of information that is now available on these topics is of astronomical proportions. Thus, the problem of deciding what to leave out has become as difficult as the decision of what to include. The guiding principle in writing this set has always been to provide a thorough overview of the topics without overwhelming the reader with a mountain of facts and figures. To be sure, this set contains many facts and figures, but these have been carefully chosen to illustrate only the essential principles.

This set, in keeping with the expansion of the biological disciplines, has grown to accommodate new material and new areas of research. Four new books have been added that focus on areas of biological research that are reaping the benefits of genome science and modern research technologies. Thus, the New Biology set now consists of the following 10 volumes:

1. *Aging, Revised Edition*
2. *Animal Cloning, Revised Edition*
3. *Cancer, Revised Edition*
4. *The Cell, Revised Edition*
5. *Gene Therapy, Revised Edition*
6. *Stem Cell Research, Revised Edition*
7. *Genome Research*
8. *The Immune System*
9. *Modern Medicine*
10. *Viruses*

Many new chapters have been added to each of the original six volumes, and the remaining chapters have been extensively revised and updated. The number of figures and photos in each book has increased significantly, and all are now rendered in full color. The new volumes, following the same format as the originals, greatly expand the scope of the New Biology set and serve to emphasize the fact that these technologies are not just about finding cures for diseases but are helping scientists understand a wide range of biological processes. Even a partial list of these revelations is impressive: detailed information on every gene and every protein that is needed to build a human being; eventual identification of all cancer genes, stem cell–specific genes, and longevity genes; mapping of safe chromosomal insertion sites for gene therapy; and the identification of genes that control the growth of the human brain, the development of speech, and the maintenance of mental stability. In a stunning achievement, genome researchers have been able to trace the exact route our human ancestors used to emigrate from Africa nearly 65,000 years ago and even to estimate the number of individuals who made up the original group.

In addition to the accelerating pace of discovery, the new biology has made great strides in resolving past mistakes and failures. The Gelsinger trial was a dismal failure that killed a young man in the prime of his life, but gene therapy trials in the next 10 years will be astonishing, both for their success and for their safety. For the past 50 years, cancer researchers have been caught in a desperate struggle as they tried to control the growth and spread of deadly tumors, but many scientists are now confident that cancer will be eliminated by 2020. Viruses, such as HIV or the flu, are resourceful and often deadly adversaries, but genome researchers are about to put the fight on more rational grounds as detailed information is obtained about viral genes, viral life cycles, and viruses' uncanny ability to evade or cripple the human immune system.

These struggles and more are covered in this edition of the New Biology set. I hope the discourse will serve to illustrate both the power of science and the near superhuman effort that has gone into the creation and validation of these technologies.

# Acknowledgments

I would first like to thank the legions of science graduate students and postdoctoral fellows who have made the new biology a practical reality. They are the unsung heroes of this discipline. The clarity and accuracy of the initial manuscript for this book was much improved by reviews and comments from Diana Dowsley, Michael Panno, Rebecca and Peter Lapres, and later by Frank Darmstadt, executive editor. I am also indebted to Diane French and Elizabeth Oakes for their help in locating photographs for the New Biology set. Finally, as always, I would like to thank my wife and daughter for keeping the ship on an even keel.

# Introduction

Viruses are the most enigmatic creatures on Earth. Although they have been called living crystals, the question as to whether these things are alive or not is controversial. This fact alone complicates any discussion regarding their basic nature. Viruses certainly are not cells, although some of the larger ones have a surprisingly complex cell-like structure. This is suggestive, but some researchers believe it is not enough to justify referring to them as a life-form. A common practice, based on the notion that viruses are not alive, is to call them entities, agents, or particles.

Many scientists believe that viruses occupy a gray zone between the living and the nonliving. This idea has persisted for so long that a serious study of viral ecology has only been possible in recent years. Scientists who believe viruses are not truly alive argue that they simply acquire lifelike properties from their hosts, most notably their ability to reproduce and their highly organized structure. By extending this argument, one could assume that if all bacteria, plants, and animals disappeared suddenly from the face of the Earth that the viruses would soon follow. On the other hand, if all the viruses disappeared everything else would continue on as though nothing had happened.

Those who believe that viruses are alive counter that the above argument does not apply because it is true of all parasites. Much of the controversy has focused on the apparent simplicity of viruses, and many critics believe that such simple structures cannot really be alive. But the largest known virus, the mimivirus, which infects amoebae, is almost as complex as the simplest bacterium. Part of the problem may be that the definition of life is too rigid. As the British playwright George Bernard Shaw once wrote, "You think that life is nothing but not being stone dead."

Another aspect of a virus's enigmatic nature is the fact that they are essentially molecular creatures, often possessing no more than a dozen genes, yet are capable of surprisingly complex and cunning behavior. These tiny microorganisms have not only outsmarted the human immune system, possibly the best in the world, but thousands of scientists all over the world have struggled for nearly 30 years to control the human immunodeficiency virus (HIV), structurally one of the simplest of all the viruses. Scientists who have noted these facts believe that viruses are living organisms, possibly at the very threshold of becoming a primitive cellular life-form.

Despite the controversy, everyone agrees that viruses are nearly synonymous with sickness and disease. Viral diseases are so common we know them by name: AIDS, influenza, measles, polio, rabies, smallpox, chicken pox, and yellow fever. These are only a few of the diseases that viruses are known to cause, and if we were to estimate the number of people who have died from viral diseases over the past 200 years it would exceed the mortalities caused by wars and all other diseases combined. The U.S. Civil War, for example, claimed 650,000 lives, but most of those deaths were due to viral infections.

And yet, these masters of death and disease have also done some good: They have helped shape our ecosystems and our genomes; they have accelerated the rate at which all organisms have evolved; and they have helped scientists discover the genes that cause cancer. Biotechnology, gene therapy, and some forms of stem cell therapy all depend on viruses, and together those procedures are being used to find cures for cancer, cardiovascular disease, neurological disorders, and many other diseases. Virus-dependent research and medical therapies may eventually save millions of lives every year. Thus, while viruses have been our deadliest foes, they have also been valuable allies.

*Viruses,* one volume in the New Biology set, covers the structure, function, and evolution of viruses with an emphasis on their dual role as infectious microorganisms and important members of Earth's biosphere. Viruses appeared not long after the evolution of the first cells and have played an extremely important role in Earth's ecology, a fact that scientists have only recently begun to appreciate.

The first four chapters of this book discuss the origin of viruses, viral structure and behavior, viral taxonomy, and the history of virology. Viral taxonomy is somewhat controversial and is currently in a state of flux. The most comprehensive scheme is presented here. Subsequent chapters are devoted to discussions of marine virology, the use of viruses in biomedical research, viral diseases, and modern methods for fighting viral infections. Viral diseases have been studied for more than 200 years, but marine virology is a relatively new discipline that has provided some startling insights into the role viruses play in the ecology of the sea. One chapter is devoted to viral pandemics, a topic that the general public has recently become interested in as health agencies, such as the World Health Organization (WHO), release annual bulletins regarding the threat of possibly deadly viral strains. The final chapter provides background material on cell biology, biotechnology, methods in virology, and other relevant topics.

# The Origin of Viruses

Fifty years ago, scientists thought that viruses were examples of the most primitive of all life-forms, giving rise to prokaryotes (bacteria) and eukaryotes (cells that formed plants and animals). Today, researchers know this is not true. Viruses appeared after the emergence of prokaryotes, the most ancient form of cell, and it is believed they evolved from mobile genetic elements known as transposons and bacterial minichromosomes known as plasmids. Because of their small size, transposons and plasmids could "leak" out of a bacterial cell after which they could enter, or infect, other cells. Over time these genetic elements acquired a protein coat that protected their DNA and made it easier for them to infect cells. It is thought that transposons evolved into eukaryote viruses while plasmids evolved into prokaryote viruses. Some of the ancestral eukaryote viruses simply took up residence in the genome where they were able to move from one chromosome to another, but stayed within the nucleus. Other ancestral viruses, however, went on to evolve into complex viruses that could leave a cell, often destroying it in the process, and could then reinfect others.

This chapter discusses the origin of viruses within the context of the origin of the cell. The first cells had many problems to solve before they could flourish in what was almost certainly a hostile environment. Perhaps the biggest problem to overcome was the establishment of some form of inheritance system long before genes and sexual reproduction had evolved. The first cells accomplished this simply by swapping molecules with their neighbors. But this process, so essential for the emergence of life, set the stage for the appearance of the first viruses.

## THE FIRST CELLS SET THE STAGE

Life appeared on Earth for the first time as single cells about 3.5 billion years ago. At that time Earth was a hot and stormy planet with surface temperatures exceeding 150° F (65° C), and an atmosphere that consisted primarily of methane and ammonia. The violent electric storms that were common in those days were crucial for the origin of life for they provided the necessary energy for the synthesis of organic compounds from the methane and ammonia. Scientists have shown that the organic compounds so produced included amino acids, nucleic acids, sugars, and fats, all of which are essential ingredients in living things today. The heat fused many of these molecules into macromolecules (chains of molecules) such as protein, deoxyribonucleic acid (DNA), ribonucleic acid (RNA), and a fatty substance called phospholipid. These compounds enriched the water, turning the oceans into a nutrient broth. The phospholipid, unlike the other macromolecules, could not dissolve in water but spread out on the surface, producing Earth's first oil slick (for more detail see chapter 10).

In addition to the heat and electrical discharges, the storms provided something else that was essential for the appearance of

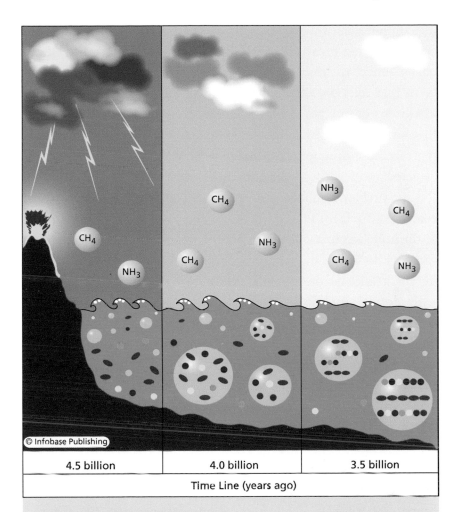

| 4.5 billion | 4.0 billion | 3.5 billion |
|:-----------:|:-----------:|:-----------:|

Time Line (years ago)

The origin of the first cells. Organic molecules essential for life were synthesized spontaneously 4.5 billion years ago when Earth was hot, stormy and wracked with constant volcanic eruptions. Some of the organic molecules were captured by lipid bubbles (light blue spheres) formed by ocean turbulence near a shoreline, and by 3.5 million years ago the first cells learned how to assemble the molecules into a variety of polymers. Nucleic acids, amino acids, fats, and sugars were among the organic molecules produced in the prebiotic oceans; only the nucleic acids (colored circles) and amino acids (brown ovals) are shown. Major gases in the atmosphere included methane ($CH_4$ and ammonia ($NH_3$).

life: thunderous waves, breaking on the shores. Anyone who has watched a wave break on a shore has witnessed one of the most important mechanisms for the formation of life on this planet. The foam that rolls onto the shore after the wave breaks is composed of billions of bubbles. In the prebiotic coastal waters, each bubble that formed collected a different sample of the water and, therefore, represented a unique individual, a separate experiment that could be acted upon by the forces of natural selection.

The prebiotic bubbles were stabilized by the phospholipids that coated the surface of the water. But when a new wave arrived, many of the bubbles burst open from the turbulence, thus releasing their contents back into the environment. Over time, evolution selected for stable bubbles that spent longer periods of time experimenting with captured molecules. Presumably, this happened when a bubble managed to synthesize a compound that increased the stability of its lipid membrane. As a consequence, that bubble not only enjoyed a longer existence but the fruit of its labors, the membrane-stabilizing molecule, was eventually made available to the community as a whole. With billions of bubbles being formed, it is conceivable that within the population many other useful molecules could have been produced and then eventually released into the environment. When new bubbles formed, they may have captured some, or all, of those molecules and thus were given a head start. This simple form of genetic inheritance, acted upon by natural selection, likely transformed the prebiotic bubbles into the first cells, but it also paved the way for the appearance of the first viruses.

Two types of cells have evolved on Earth, and they in turn have given rise to different types of viruses. The first type of cell is called a prokaryote (meaning "before the nucleus"). These cells have the simplest structure and have diverged into the archaea and the bacteria. The second type of cell is called a eukaryote (meaning "true nucleus"). These cells evolved from the prokaryotes and are the type

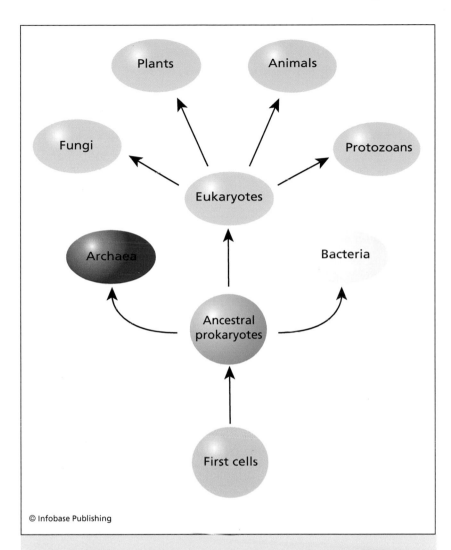

Cell classification. The first cells evolved into the ancestral prokary-otes, which gave rise to the archaea, bacteria, and eukaryotes, the three major divisions of life in the world. The archaea and bacteria are very similar anatomically but differ biochemically. Eukaryotes, anatomically and biochemically distinct from both the archaea and bacteria, gave rise to plants, animals, protozoans, and fungi.

of cell that all plants and animals are made from. There are also single-cell eukaryotes known as protozoans. Although the main focus of this book is on animal viruses, there are many different kinds of viruses that only infect prokaryotes and plants; these will be discussed in chapter 3.

Prokaryotes have a simple structure that includes a cell membrane, a protoplasm that contains the cell's DNA genome and all of the biochemical machinery that the cell needs for reproduction, energy metabolism, and the acquisition of nutrient molecules. Many of these cells also have a secondary genome, or minichromosome, known as a plasmid. This important structure is discussed in a following section. Prokaryotes, particularly the bacteria, inhabit nearly every niche on Earth, including the soil, air, and water. There are also many different species that live on the skin or in the bodies of animals, the latter of which are confined to the mouth, throat, and digestive tract.

Eukaryotes are much bigger and much more complex than prokaryotes. In a prokaryote, all of the cell's machinery and all of its biochemical activity take place in a single compartment, the protoplasm. By contrast, eukaryotes have special compartments, or organelles, for everything. The DNA is kept in the nucleus, and proteins are synthesized in the cytoplasm, some of which are modified in special organelles known as the Golgi complex and the endoplasmic reticulum. Energy is produced by the mitochondria, waste material is recycled in lysosomes, and noxious compounds are detoxified in peroxisomes.

The earliest cells to appear on Earth were submerged in an ocean full of nutrients that they could easily obtain by simple absorption. But as the cell population increased two things began to happen: The availability of nutrients began to decrease and the complexity of the nutrients began to increase. The more complex nutrients could not be obtained by simple absorption. The increasing complexity of the nutrients was due to the fact that

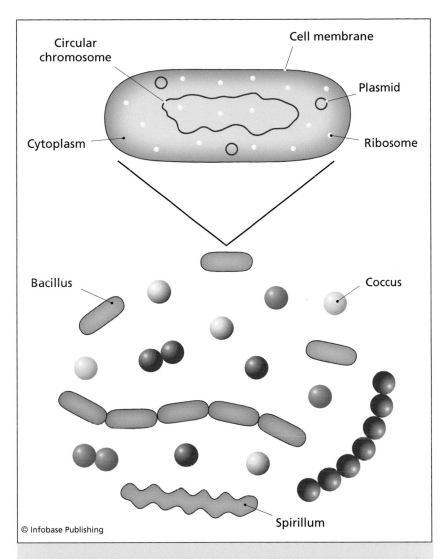

Prokaryotes. All prokaryotes have the same basic anatomy consisting of a cell membrane, a cytoplasm, and a circular DNA chromosome. Some bacteria have a second, smaller chromosome called a plasmid, which may be present in multiple copies. The cytoplasm contains a wide assortment of enzymes and molecules, as well as ribosomes, protein-RNA complexes that are involved in protein synthesis. The cells may be spherical (coccus), rod shaped (bacillus), or wavy corkscrews (spirillum), appearing singly, in pairs, or linked together into short chains.

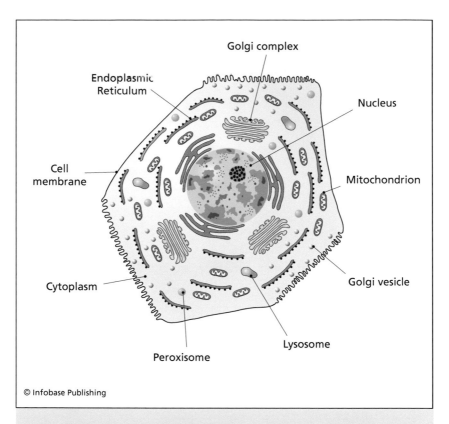

Golgi complex
Endoplasmic Reticulum
Nucleus
Cell membrane
Mitochondrion
Cytoplasm
Golgi vesicle
Peroxisome
Lysosome

© Infobase Publishing

The eukaryote cell. The structural components shown here are present in organisms as diverse as protozoans, plants, and animals. The nucleus contains the DNA genome, and as assembly plant for ribosomal subunits (the nucleolus). The endoplasmic reticulum (ER) and the Golgi work together to modify proteins, most of which are destined for the cell membrane. These proteins are sent to the membrane in Golgi vesicles. Mitochondria provide the cell with energy in the form of adenosine triphosphate (ATP). Ribosomes, some of which are attached to the ER, synthesize proteins. Lysosomes and peroxisomes recycle cellular material and molecules. The microtubules and centrosome form the spindle apparatus for moving chromosomes to the daughter cells during cell division. Actin filaments and a weblike structure consisting of intermediate filaments (not shown) form the cytoskeleton.

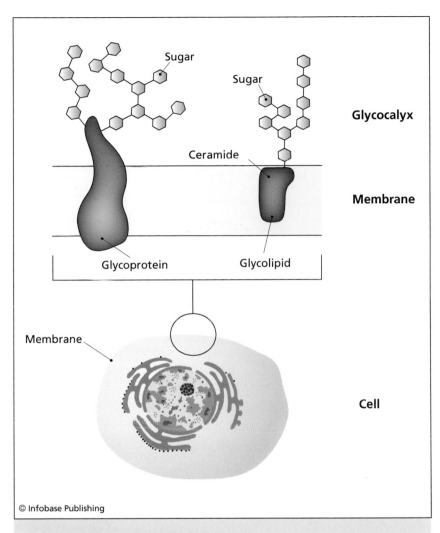

The eukaryote glycocalyx. The eukaryotes molecular forest consists of glycoproteins and glycolipids. Two examples are shown at the top, a glycoprotein on the left and a glycolipid on the right. The glycoprotein trees have "trunks" made of protein and "leaves" made of sugar molecules. Glycolipids also have "leaves" made of sugar molecules, but the "trunks" are a fatty compound called ceramide that is completely submerged within the plane of the membrane. The glycocalyx has many jobs, including cell-to-cell communication and the transport and detection of food molecules. It also provides recognition markers so the immune system can detect foreign cells.

they were coming from cells that had died, rather than the pool of simple prebiotic nutrients. In addition, by this time many cells had learned how to capture energy from the Sun. These were the

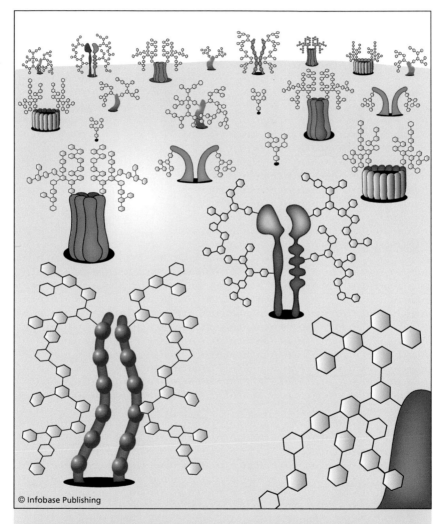

© Infobase Publishing

A panoramic view of the glycocalyx. The glycoproteins in the cells's forest come in many different shapes and sizes, and they dominate the surface of most cells. The glycolipids all have the same ceramide trunks, but the molecular foliage varies considerably. All but three of the structures in this image are glycoproteins, but in nerve cells glycolipids are much more common.

first autotrophs, cells that could photosynthesize, and with that ability came an ever-increasing population of complex macromolecules that were released into the water when those cells died. Consequently, the eukaryotes, like the prokaryotes before them, developed cell-surface receptors that could capture and ingest large molecules. The receptors are glycoproteins (proteins with sugar molecules attached) that are part of a forest of macromolecules called the glycocalyx that cover the surface of the cell. This structure also gives the cell its ability to communicate with other cells. Viruses have learned how to exploit the normal function of the glycocalyx in order to gain access to the cell.

Transporters, and their associated receptors, are the specific glycocalyx structures used by viruses to enter a cell. These structures consist of several glycoproteins arranged in a porelike configuration and are in effect doorways into the cell. Some of these receptors are associated with a process called endocytosis whereby large macromolecules and even whole cells are brought into the cell by the invagination of the cell membrane. Viruses have discovered the keys to these ports of entry, and they use them every time they infect a cell. The details of these transporters, as they pertain to viral infections, will be discussed in a later chapter.

## MOBILE GENETIC ELEMENTS

The first cells are believed to have had a small RNA genome that encoded fewer than a dozen genes. (Modern cells all have DNA genomes.) When these cells died, they released their genome into the water along with everything else. Because their genomes were so small, it is possible that other cells could have picked them up in their entirety. The host cell could have kept them as a second chromosome, but after a period of time biochemical systems evolved that incorporated the captured genes into the host's genome. Thus, early cells were not only involved in swapping useful construction materials but were also involved in swapping genes. Once captured, some of these genes were able to move around from one location in the genome to another. Scientists believe that these genetic elements gave rise to viruses that infect eukaryotes.

When first proposed by the American geneticist Barbara McClintock in 1951, the idea that genes could move from one location in the genome to some other location was greeted with disbelief and disdain. For more than 20 years, this idea was dismissed as wild speculation until the advent of recombinant technology made it possible to prove the existence of these wandering genes, also known as transposons, transposable elements, and jumping genes. McClintock's work was eventually given the recognition it deserved, and in 1983, at the age of 81, she was awarded the Nobel Prize in physiology or medicine. She died on September 2, 1992.

McClintock's research provided the first clues into the origin of viruses, particularly those that infect animal cells. Somehow, millions of years ago, a jumping gene learned how to jump right out of the cell (that is, the genetic element did not have to wait for the cell to die before it could leave). It acquired this ability in small steps as it moved from one place in the genome to another. A jumping gene could conceivably reinsert next to a gene that codes for a potential capsid protein (a protein that surrounds and protects a virus's genome). In effect, the capsid gave the virus a body, with structure and form. The next time such a transposon moved it could take a copy of the potential capsid gene with it. Eventually, by moving from place to place, the transposon would have collected a large number of genes that not only made it possible for it to escape from the cell but also gave it the power to reinfect other cells. When that happened, a simple mobile genetic element went from being a molecular curiosity to a living thing, equipped with a life cycle and the power of reproduction.

Viruses, and the transposons they evolved from, have had a profound effect on the evolution of the animal genome. For example, only about 2 percent of the human genome contains genes, with the rest of the DNA consisting of intervening sequences. These are gene-free areas of the genome that may be thought of as safe zones, within which a virus or a transposon may settle without causing any damage. If a virus does happen to insert within a gene, the resulting damage, known as insertional mutagenesis, can have serious consequences and is known to be the cause of certain forms of cancer. In order to minimize this problem, the human genome, and indeed all animal genomes, have evolved into a form that accommodates mobile genes. In such a genome, the odds of a virus or a transposable element damaging an existing gene are extremely small.

Transposons move by being replicated (duplicated) by nuclear enzymes. Thus, the original transposon stays put, while the daughter strand moves to a new location. Consequently, over time the genome becomes sprinkled with many copies of transposable elements. These copies of the original element are then free to mutate into genes that may eventually become useful to the organism. Thus, transposons are the source of many of the genes now present in the human genome. When certain viruses infect a cell, they too take up residence within an intervening sequence. Recent studies have shown that the human genome is littered with viral genomes. Fortunately, most of these have been inactivated by slow mutational changes and are effectively "locked in" to the genome.

The flexibility of a transposable genome is perhaps the single most important characteristic that led to the explosive adaptability of eukaryotes and the many life-forms they produced. Such a genome is also important to modern medical therapies, such as gene therapy, which attempts to cure a disease by introducing a normal gene into the patient's genome. If the human genome were organized like that of the prokaryotes, which lacks intervening sequences, such a therapy would be nearly impossible.

## PLASMIDS

A plasmid is a special type of mobile genetic element that occurs only in prokaryotes. These genetic elements are never integrated into the host chromosome (owing to the lack of "safe zones") but serve instead as independent mini-chromosomes. It is likely that bacteria have been exchanging plasmids among themselves for more than a billion years. This ancient practice was retained because it is often of mutual benefit. Plasmids carry antibiotic-resistance genes, so if a bacterium happens to make one that is very effective, an unrelated bacterium could get a copy simply by capturing the plasmid. Plasmids were probably released into the environment when a cell's membrane became leaky, for various reasons, or when the cell died and broke open, an event that echoes the molecular swapping discussed above. But plasmid exchange among prokaryotes could only work as long as the plasmids stayed small enough to reenter an intact cell by passive diffusion.

The first bacterial virus, known as a bacteriophage or phage, was probably a plasmid that picked up a gene for a protein that could spontaneously form a capsid. Acquiring a capsid made it possible for the phage to interact with bacterial cell-surface receptors so the virus was no longer dependent on passive diffusion for entry. Once the cell-surface barrier was overcome, the phage genome was free to increase in size from a few genes to a more than a dozen. With a larger genome, the bacteriophage developed a number of strategies for infecting cells and commandeering the cellular machinery to suit their own needs.

## SUMMARY

Viruses arose as a direct consequence of the special requirements for the origin and proliferation of the first cells. Swapping molecules and the evolution of the glycocalyx were essential for the success of the first cells. But having established the sharing of genetic material as a way of life, it was only a matter of time before some of those genes evolved into infectious microorganisms. Scientists believe that plant and animal viruses evolved from transposable elements and that prokaryote viruses evolved from plasmids.

# Viral Structure
# and Behavior

Soon after James Watson and Francis Crick resolved the structure of DNA, Watson published a paper on viral structure in which he suggested that since a virus is such a tiny particle, less than one-tenth the size of a bacterium, it could only carry enough nucleic acid for a dozen or so genes. Consequently, he proposed that viral structure must consist of only a few proteins used over and over again in some sort of symmetrical highly ordered arrangement. To test this idea, many biologists examined viral structure under the newly available electron microscope, and when they did they saw tiny crystalline structures that confirmed Watson's speculations.

Most viruses have a crystalline protein structure that is icosahedral (constructed from triangles, like a geodesic dome). The protein crystal forms a hollow compartment called the capsid that contains the viral genome. In some cases an envelope consisting of a lipid bilayer, which is often studded with proteins, surrounds the crystalline capsid. Some viruses, such as the influenza virus, have a simple though highly ordered spherical capsid instead of a crystalline icosahedron. There are essentially three different types of viruses: DNA viruses (i.e., they have a DNA genome), RNA viruses, and retroviruses. The third type of virus, the retrovirus, is a special kind of RNA virus that converts its genome to DNA after infecting the host. Additional features distinguish viruses within these main groups: the presence or absence of an envelope, the structure of the capsid, and some detailed features concerning the genome; that is, whether the DNA is linear or circular and how many genes it encodes. Examples discussed in this chapter are animal viruses.

## DNA VIRUSES

The adenovirus, which infects the upper respiratory tract in humans, is an example of a DNA virus. The genome is housed in an icosahedral capsid, which not only protects the genome but also carries the "keys" that the virus needs to gain access to a cell.

Owing to the geometry of the icosahedron, these capsids have 12 vertices, each of which sprouts a long protein spike. The tip of these spikes contains a protein that recognizes and binds to the cell-surface receptors. The capsid consists of 11 distinct proteins arranged as penton and hexon capsomeres (clusters of identical proteins), with the spikes projecting from the pentons. The adenovirus, like all viruses, is extremely small with a diameter of about 80 nm. To put this into perspective, 80 nm is equal to 0.08 μm. A bacterium has a diameter of about 1.0 μm and a eukaryote is usually 10 to 20 times larger. If the adenovirus were the size of a period on this page, the bacterium would have a diameter of about half an inch (10.0 mm) and the eukaryote would have a diameter of about 4.0 inches (100 mm).

All cells, whether they are prokaryotes or eukaryotes, have DNA genomes. DNA is a very stable molecule that can store many thousands of genes, essential for the complex lifestyles of modern

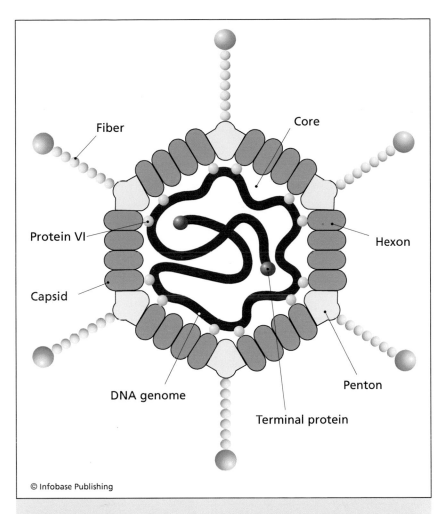

Fiber

Core

Protein VI

Hexon

Capsid

DNA genome

Penton

Terminal protein

© Infobase Publishing

Structure of the adenovirus. The capsid is constructed from repeating hexon and penton capsomeres. A long fibrous protein, attached to each penton, is crucial for cell entry. The DNA chromosome, anchored by protein VI, contains 30 to 40 genes and is stabilized by two terminal proteins. Several other proteins, not shown, are stored in the core to initiate and maintain infection.

cells, which usually have 4,000 to 30,000 genes. In addition, double-stranded DNA allows for error correction, an extremely important feature when millions of nucleotides are to be replicated.

Viruses, on the other hand, are simple microorganisms, so simple that many of them get by with fewer than a dozen genes. When genomes are this small, the advantage of DNA over RNA disappears. Consequently, the viral world is almost evenly split between those that have a DNA genome and those that use RNA instead. There is, however, a great deal of variation on this theme. DNA ge-

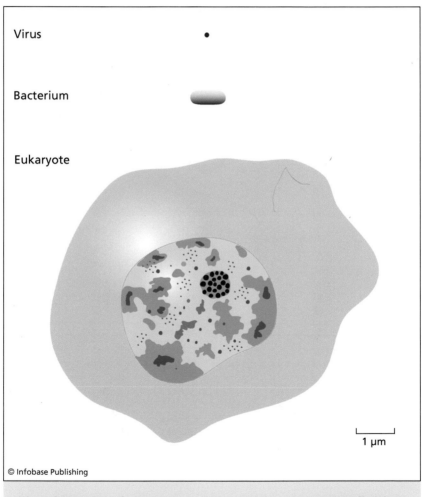

Virus

Bacterium

Eukaryote

1 μm

© Infobase Publishing

The relative size of a virus. A virus, top panel, is 10 times smaller than a bacterium, which is 10 to 20 times smaller than a eukaryote cell. The size bar is 1.0 μm.

nomes may be single-stranded or double- , circular or linear, and they may appear as a single piece of DNA or two or more pieces. RNA genomes are always linear, consisting of one or more strands. In addition, the RNA may be a positive (+) strand, meaning that it can serve directly as a messenger RNA (mRNA), or a negative (-) strand, which is converted to a positive strand during the infection cycle. Viruses with only a few genes have a single-stranded RNA genome, but as the number of genes increases, the genome tends to be double-stranded DNA.

The adenovirus has a double-stranded linear DNA chromosome, containing 30 to 40 genes, that is capped at each end with a terminal protein offering added stability to the molecule. More than half of the genes code for capsid proteins and enzymes that are needed for infection. A few of the genes code for histone-like proteins that bind to the chromosome for added stability. Adenoviruses, so named because they were originally isolated from the adenoid glands, are responsible for general infections of the upper respiratory tract.

All cells need to communicate with the outside world, and in particular they need a system for detecting and collecting nutrients. Prokaryotes satisfied these requirements by embedding protein receptors in their cell membrane. These receptors are part of the glycocalyx, a molecular forest that covers the cell membrane. With the establishment of the glycocalyx, the cell membrane became increasingly resistant to the passive diffusion of large molecules. That is, it tended not to leak as much as it did in primitive cells. This trend, begun by the prokaryotes, was converted to a rule by the eukaryotes: Anything coming in has to pass a port of entry.

Ports of entry include sugar receptors, hormone receptors, and ion channels. Ion channels are usually open, permitting free access to the cell's interior, but the pore size is small enough to block entry to large molecules and microbes. The sugar receptors and some hormone receptors are linked to a process called endocytosis that is able to bring large molecules and microbes into the cell. When

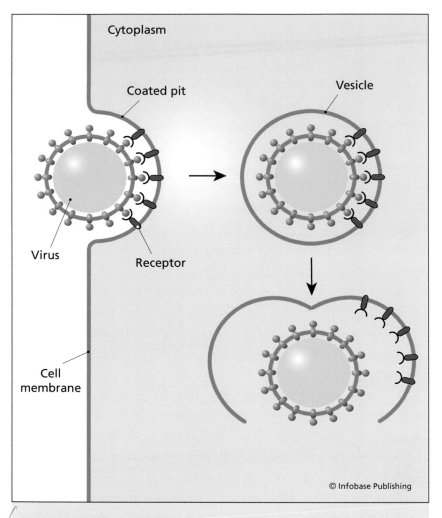

Receptor-mediated endocytosis. A virus enters a cell by binding to receptors in a coated pit, which activates endocytosis; a process the cell normally uses to ingest food or signaling molecules. Once inside, a viral enzyme attacks the wall of the vesicle, causing it to rupture, releasing the virus into the cytoplasm.

the proper molecule makes contact with these receptors, it is drawn inside by the formation of a vesicle (bubble). This type of entry is called receptor-mediated endocytosis.

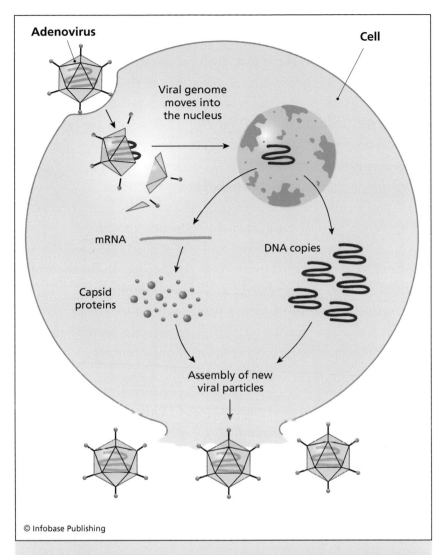

Life cycle of a DNA virus (adenovirus). After the virus enters the cell, the fragmented capsid docks at a nuclear pore (not shown) and releases the chromosome into the nucleus, where it is replicated and transcribed. The replicated DNA and the mRNA leave the nucleus and enter the cytoplasm. The viral mRNA is translated in the cytoplasm, and the proteins join with the DNA copies to form new viral particles, which leave the cell by disrupting the membrane.

Viral capsids can activate endocytosis to gain entry into a cell. Once inside, the virus releases an enzyme that attacks the wall of the vesicle, causing it to rupture, thus releasing the viral capsid into the cytoplasm. There are several variations to this scheme: For example, the AIDS virus binds to cell receptors but does not activate endocytosis. Instead, the viral envelope fuses with the cell membrane, releasing the capsid directly into the cytoplasm without the formation of a vesicle. In all cases, once the capsid is free in the cytoplasm, it breaks open to release the viral genome.

When an adenovirus enters a cell, the partially fragmented capsid binds to a nuclear pore after which the chromosome moves into the nucleus where it is replicated and transcribed. Viral mRNA and copies of its genome then move from the nucleus to the cytoplasm where the mRNA is translated into capsid and other viral proteins. The replicated genome and the newly synthesized viral proteins auto-assemble into mature viral particles, which leave the cell by rupturing its membrane, killing the cell in the process.

## RNA VIRUSES

The influenza virus is a typical example of an RNA virus. The genome consists of eight pieces of single-stranded RNA, all of which are negative strands. Illustrations usually depict the genome as a pyramid, with the shortest strand at the top and the longest strand at the bottom. In reality, proteins associated with the RNA twist the eight strands into a complex helical structure, somewhat like an eight-stranded piece of conical rope. The genome codes for nine proteins, most of which are structural. This virus has a spherical capsid that is surrounded by a lipid envelope, derived from the cellular host. The envelope contains two viral-encoded glycoproteins. The first of these is called hemagglutinin (HA) and the second is called neuraminidase (NA). The HA glycoprotein is the most prevalent, accounting for about 25 percent of viral protein, whereas the NA glycoprotein accounts for about 5 percent. Several subtypes are

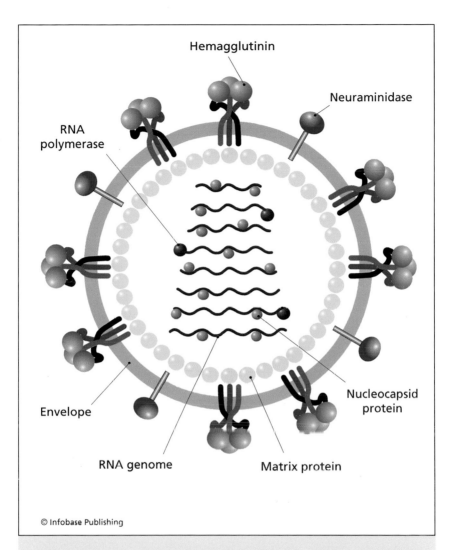

Influenza virus. This virus has a spherical capsid that is surrounded by a lipid envelope, derived from the cellular host. The envelope contains two viral-encoded glycoproteins: hemagglutinin and neuraminidase. The RNA genome consists of eight chromosomes that are associated with nucleocapsid proteins and RNA polymerase. Although some of the elements in the figure are shown in three dimensions, the illustration is primarily a two-dimensional slice of a spherical structure.

known to exist for each of these glycoproteins (e.g., H1, H2, N1, N2). These viral surface antigens are the source of the current naming convention of influenza viruses, the most recent being the H1N1 or swine flu virus.

Flu viruses enter the cell by receptor-mediated endocytosis. The vesicle ruptures, releasing the RNA genome and the associated proteins into the cytoplasm. The core proteins remain attached to the RNA, and together they migrate into the nucleus. Once they are in the nucleus, the nucleocapsid proteins dissociate and the viral RNA polymerase synthesizes viral mRNA and the viral genome,

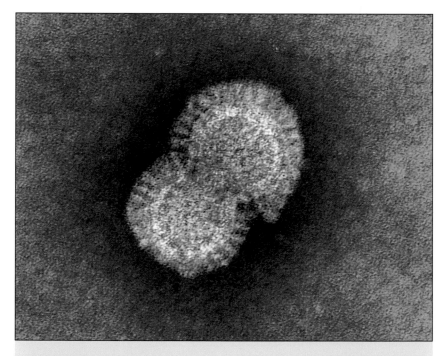

Avian influenza. A transmission electron micrograph (TEM) of the H5N1 avian flu virus, a member of the Orthomyxoviridae family. These viruses cause an infectious and contagious respiratory disease that often results in a pandemic or smaller seasonal epidemics. *(World Health Organization)*

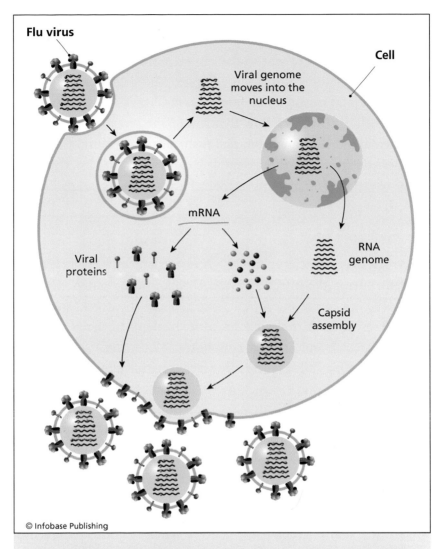

**Flu virus**

Cell

Viral genome moves into the nucleus

mRNA

RNA genome

Viral proteins

Capsid assembly

© Infobase Publishing

Life cycle of an RNA virus (influenza). Flu viruses enter the cell by receptor-mediated endocytosis. The RNA genome translocates to the nucleus where it is transcribed and replicated. The replicated viral genome and the mRNA leave the nucleus and enter the cytoplasm. The viral mRNA is translated in the cytoplasm, and the proteins eventually join with the viral genome to form new viral particles. The daughter virions acquire an envelope as they leave the cell by exocytosis.

both of which translocate to the cytoplasm. The mRNA is translated into capsid and other viral proteins. The glycoproteins, destined for the viral envelope, are sent to the cell membrane by way of the endoplasmic reticulum and Golgi complex. These cellular organelles glycosylate (add sugars to) certain proteins. Thus, any glycoprotein, viral or cellular, has to follow that route. The rest of the viral proteins remain in the cytoplasm and form a capsid around the newly synthesized genome. The assembled capsid obtains an envelope, made of cell membrane and retroviral glycoproteins, while leaving the cell by exocytosis.

## RETROVIRUSES

A retrovirus is a special type of RNA virus that converts its genome to DNA during the infection cycle. Human immunodeficiency virus (HIV) is a common example. The genome consists of two identical pieces of RNA that encodes nine overlapping genes, three of which (*Gag, Pol,* and *Env*) are common to all retroviruses. HIV has a special enzyme called reverse transcriptase that converts the RNA genome to DNA after it infects a cell. This enzyme allows the virus to reverse the usual DNA to RNA direction of genetic biosynthesis and is the reason they are called retroviruses. The HIV capsid consists of a spherical protein matrix immediately beneath the envelope and a cone-shaped core that forms the genome compartment. The proteins embedded in the envelope are of viral origin and are essential for entering the cell. HIV has a diameter of about 100 nm.

An unusual characteristic of RNA viruses is the presence of overlapping genes. In such a genome, one stretch of the chromosome can be used to code for two or three genes. These viruses also process their mRNA to produce more than one protein from any given gene. When this occurs, a precursor mRNA is synthesized from the gene, after which it is split in half to produce two different proteins. Thus, the *Env* gene codes for two envelope proteins, glycoprotein 120 (gp120) and gp41. Likewise, the *Gag* gene codes for the matrix

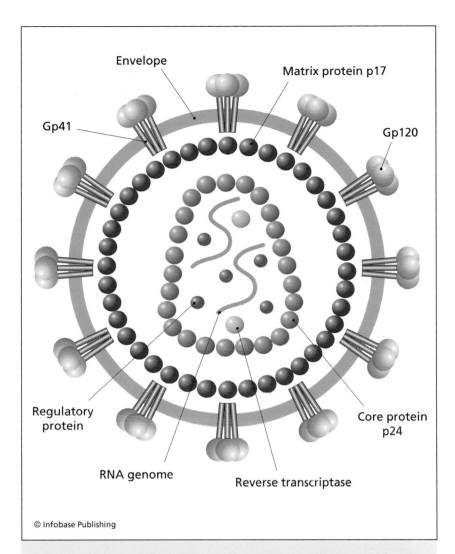

Envelope

Matrix protein p17

Gp41

Gp120

Regulatory protein

Core protein p24

RNA genome

Reverse transcriptase

© Infobase Publishing

Structure of a retrovirus (HIV). The example, shown in cross section, is HIV, an enveloped retrovirus that has a double-stranded RNA genome containing nine genes. The capsid consists of a spherical matrix and an inner, cone-shaped protein core. Glycoproteins (Gp) 41 and 120, embedded in the envelope, are crucial for cell entry. Several regulatory proteins, including reverse transcriptase, are stored in the core.

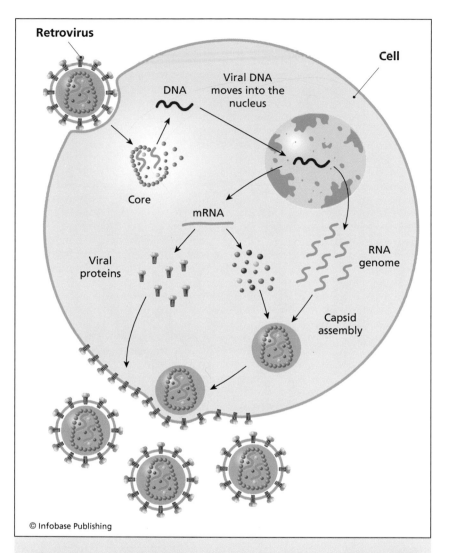

Life cycle of a retrovirus (HIV). After the virus enters the cell, the RNA chromosome is released from the core and copied into DNA by reverse transcriptase. The DNA chromosome enters the nucleus, where it integrates into a host chromosome, after which it is transcribed into RNA. The RNA leaves the nucleus, some of which is translated into capsid and envelope proteins, and the rest becomes new copies of the RNA genome. The translated viral proteins are embedded in the membrane. The assembled capsid obtains an envelope, made of cell membrane and retroviral proteins, while leaving the cell by exocytosis.

protein (p17) and the core protein (p24). The numbers refer to the relative sizes of the proteins. In addition, genes for six regulatory proteins overlap the *Env* gene region of the chromosome. The *Pol* gene codes for reverse transcriptase, which is an RNA-dependent DNA polymerase. In other words, the polymerase uses RNA as a template to synthesize a complementary strand of DNA. Despite their simple structure and small genome, retroviruses are among the most virulent pathogens known.

The life cycle of a retrovirus, such as HIV, is more complex than that of a DNA or RNA virus. When a retrovirus infects a cell, the capsid core breaks open to release the RNA genome into the cytoplasm, which is quickly converted to DNA by reverse transcriptase. The viral DNA moves into the nucleus and inserts itself into one of the cell's chromosomes. After insertion, the viral genes are transcribed, producing mRNA, and the entire length of viral DNA is transcribed to produce many copies of the viral RNA chromosome. The mRNA and the RNA chromosomes migrate back to the cytoplasm where the mRNA is translated into capsid, regulatory, and envelope proteins, the latter of which is sent to the cell membrane by way of the endoplasmic reticulum and the Golgi complex.

New capsids form by the auto-assembly of the newly made RNA chromosomes and capsid proteins. The cell membrane, now studded with viral proteins, then forms an envelope around the viral particles as they leave the cell by exocytosis. Thus, as with influenza viruses, the lipid bilayer that surrounds a retrovirus is obtained from the cell, while the envelope proteins are of viral origin. An important feature of this life cycle is that the virus does not rupture the membrane and hence does not kill the cell when it leaves. &

## SUMMARY

Viruses are the smallest microorganisms known. On average, they are only one-tenth the size of a bacterium and are usually assembled from fewer than a dozen different kinds of proteins. The "body" of a virus is called a capsid, which is a hollow structure that contains

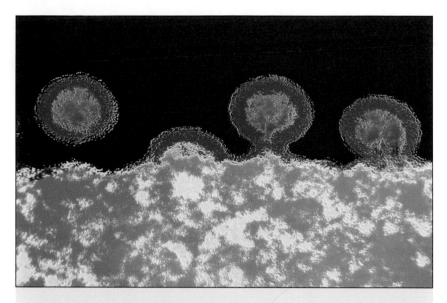

HIV viruses budding from an infected human T lymphocyte. The cell is at bottom (pink). Four viruses are seen in different stages of budding: At center left the virus acquires its coat from the cell membrane (red); at right the virus buds from the cell; at center right budding is almost complete; at left the new virus is free-floating. Once free, the HIV virus with central RNA (green) reinfects other T cells. T cells form part of the body's immune response and are weakened by the HIV virus. Magnification: 86,000×. *(NIBSC/Photo Researchers, Inc.)*

the viral genome. The capsid may have an icosahedral geometry or may assume a spherical or filamentous shape. Viruses have either a DNA or RNA genome. RNA viruses also consist of a special group, the retroviruses, that convert their genomes to DNA as part of the infection cycle. Retroviruses have the unusual ability of "burying" the DNA version of their genome within the host's DNA. The life cycles vary considerably among the various kinds of viruses. Bacteriophages always destroy the host cell as they complete their life cycle, and this is also true for adenoviruses that infect human and other animal cells. Retroviruses, on the other hand, exit the cell via exocytosis, which does not disrupt the cell membrane and thus does not kill the cell.

# Viral Taxonomy

Viral populations are extremely diverse, reflecting the fact that they infect a wide variety of cells. This diversity, and the on-going debate regarding the essential nature of a virus, has posed enormous taxonomic challenges. In addition, since viruses are parasites, taxonomists have to include host preferences into the classification scheme along with basic morphological character-istics. For example, prokaryotes and eukaryotes have their own viruses. A prokaryote virus known as lambda infects the bacte-rium *Escherichia coli* but does not infect other bacteria. Eukaryote viruses, because they infect many multicellular organisms, are both cell and species specific. The adenovirus, for example, infects humans but is normally restricted to cells in the upper respiratory tract, including the adenoids, from which this virus gets its name. The AIDS virus (HIV) is known to infect humans, monkeys, and chimpanzees but not dogs or cats. Moreover, HIV infects white blood cells and no others. The influenza virus appears as several different strains that specialize in attacking birds or mammals and only rarely is an avian strain able to infect humans or other mammals. This type of information is essential for any viral classification scheme. Although there are a few schemes available, this chapter will discuss the one produced by the International Committee on Taxonomy of Viruses (ICTV).

## VIRAL FAMILIES

The ICTV has organized more than 5,000 types of bacteriophage into 13 families and 4,000 animal and plant viruses into 56 families, 24 of which infect humans. The committee used the following characteristics for classifying the viruses.

1.  The type of genome: Whether it is DNA or RNA, the number of chromosomes, whether the genome is double-stranded (ds) or single-stranded (ss), the size of the genome (in bases or nucleotides if single-stranded, or base pairs [bp] if double-stranded), whether it is linear or circular, and in the case of an RNA genome, whether it is a plus (+) or minus (-) strand. The proportion of guanine and cytosine nucleotides is also considered along with other special DNA and RNA features. Viral genomes range in size from about 5,000 to 40,000 bases, encoding six to 100 genes. By comparison, a bacterial genome is usually more than 2 million bp, encoding 4,000 genes, and the human genome is 3 billion bp, encoding 30,000 genes.

2.  Virion morphology: This includes the size, shape, type of symmetry, the presence or absence of an envelope, and the type of glycoproteins located in the envelope.

3.  Physical properties of the capsid: molecular mass, pH stability, thermal stability, and resistance to ether and detergents.

4.  Properties of the viral proteins: the number of proteins produced, the number of structural versus enzymatic proteins, and specific features of the glycoproteins.
5.  Biological properties: host range, pathogenicity, and cellular specificity.

The creation of a taxonomic scheme requires public scrutiny and debate, leading to formal approval by the full membership of the ICTV. On the other hand, naming a novel virus and assigning it to a preexisting species are not taxonomic acts and therefore do not require formal ICTV approval. In such cases, approval is obtained by publication of a paper describing the new virus in a peer-reviewed scientific journal. Thus, the universal virus taxonomy is supported by verifiable data and by the consensus of virologists.

## BACTERIAL VIRUSES

Viruses that infect bacteria are called bacteriophage or phage (from "bacteria" and the Greek *phagin* "to eat"). Bacteriophages infect more than 140 bacterial genera that live in the soil, freshwater, ocean environments, and the intestinal tracts of virtually all animals. Viral populations are known to be especially dense in the ocean where more than 70 percent of marine bacteria are infected. It is not uncommon for a single drop of ocean water to contain nearly a billion virions.

Most bacteriophages are DNA viruses; only two of the 13 families have RNA genomes. The genome may be circular, linear, single-stranded, or double-stranded. The capsid is usually icosahedral, with or without a hollow tail, and ranges in size from 35 to 65 nm. The exact geometry of the icosahedron is either isometric (i.e., all surfaces of equal size) or prolate (lengthened along the polar axis). The tails are quite variable in length, ranging from 10 to 250 nm long. Bacteriophages usually do not have envelopes.

The best-studied bacteriophages are those that infect bacteria in the human respiratory and digestive tracts. These bacteria, known

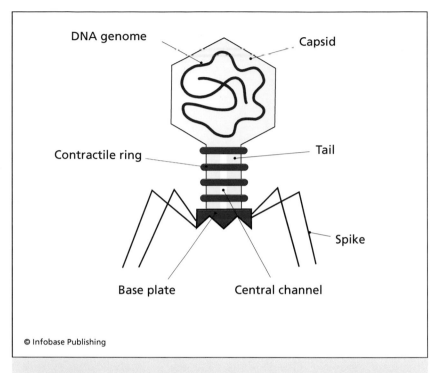

DNA genome

Capsid

Contractile ring

Tail

Spike

Base plate    Central channel

© Infobase Publishing

Bacteriophage. Most bacteriophages are DNA viruses. The genome may be circular, linear, single-stranded, or double-stranded. The naked capsid is usually icosahedral with or without a hollow tail and ranges in size from 35 to 65 nm. The exact geometry of the icosahedron is either isometric (i.e., all surfaces of equal size) or prolate (lengthened along the polar axis).

collectively as proteobacteria and enterobacteria, include such species as *E. coli., Salmonella, Proteus vulgaris,* and *Pseudomonas aeruginosa.* Many prophages infect the ancient thermophilic bacteria known as the Archaea. The relationship between these viruses and their prokaryote hosts will be discussed in chapter 5. A much-studied bacteriophage is the T2 bacteriophage *(Myoviridae)* that infects the common human intestinal bacterium *E. coli.* This phage was used by the American geneticists Alfred Hershey and Martha Chase in a classic study designed to show that DNA contains the

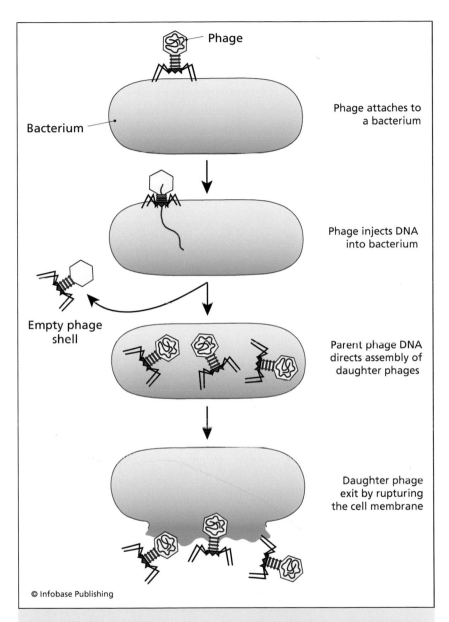

Phage

Phage attaches to
a bacterium

Bacterium

Phage injects DNA
into bacterium

Empty phage
shell

Parent phage DNA
directs assembly of
daughter phages

Daughter phage
exit by rupturing
the cell membrane

© Infobase Publishing

Bacteriophage life cycle. The phage attaches itself to the bacterial membrane and then injects its genome into the cell. The cell's biosynthetic machinery transcribes the phage genome and synthesizes all of the viral components, which auto-assemble into daughter phages. The viruses exit the cell by rupturing its membrane, killing it in the process.

cell's genes and not protein. Lambda phage *(Siphoviridae)* has a similar morphology and was used by scientists in the 1970s to develop recombinant DNA technology. The role of bacteriophages in biomedical research will be discussed in chapter 6. The T2 phage has a bare icosahedral capsid, a contractile tail, and a linear DNA genome. These viruses infect bacteria by binding to cell surface receptors after which they inject their genome into the cell.

Bacteriophages, being able to kill bacteria, have been used as an antimicrobial agent. Indeed as early as 1917, scientists at

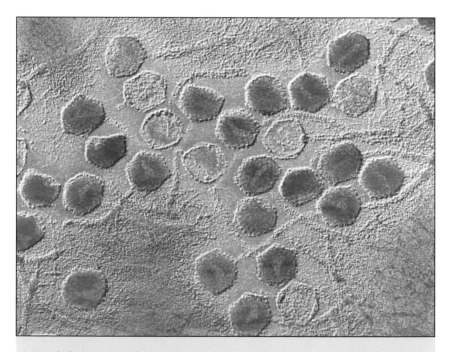

Lambda bacteriophage. The image is a false-color transmission electron micrograph (TEM). Lacking any reproductive machinery themselves, they infect bacteria and commandeer their cellular apparatus for purposes of replication. The genetic material (DNA) of the phage is contained within the head, an icosahedral figure (some colored green, others orange). Each head is attached to a (linear) tail, which in this species is noncontractile. Magnification: 18,000×. *(CNRI/Photo Researchers, Inc.)*

the Institut Pasteur had discovered a bacteriophage that could destroy the bacterium that causes dysentery. Subsequently, in limited trials, scientists tested the effectiveness of phage therapy to treat dysentery and cholera and to kill bacteria associated with food poisoning, such as *Escherichia coli* and *Salmonella*. Although much of this work was abandoned with the advent of antibacteria drugs, the FDA has recently approved using bacteriophage on cheese and other food products to kill the bacterium

## CLASSIFICATION OF BACTERIOPHAGE

| FAMILY | MORPHOLOGY | HOST | GENOME |
|---|---|---|---|
| Myoviridae | Icosahedron, contractile tail | Cyanobacteria | DNA[1] |
| Siphoviridae | Icosahedron, non-contractile tail | Proteobacteria | DNA[1] |
| Podoviridae | Icosahedron, short tail | Staphylococcus | DNA[1] |
| Tectiviridae | Icosahedron | Enterobacteria | DNA[1] |
| Corticoviridae | Icosahedron | Proteobacteria | DNA[3] |
| Lipothrixviridae | Rod-shaped, envelope | Archaea | DNA[1] |
| Plasmaviridae | Pleomorphic, envelope | Enterobacteria | DNA[3] |
| Rudiviridae | Rod-shaped | Archaea | DNA[1] |
| Fuselloviridae | Lemon-shaped, envelope | Archaea | DNA[3] |
| Inoviridae | Filamentous | Enterobacteria | DNA[4] |
| Microviridae | Icosahedron | Purple bacteria | DNA[4] |
| Leviviridae | Icosahedron | Enterobacteria | RNA[2] |
| Cystoviridae | Spherical, envelope | Pseudomonas | RNA[1] |

Note: Bacteriophage genomes are either DNA or RNA, but there is considerable variation concerning the exact form that they take. That is, the genome may be linear double-stranded (1), linear single-stranded (2), circular double-stranded (3), or circular single-stranded (4).

*Listeria monocytogenes.* In addition, some scientists believe that phage therapy may prove useful in the fight against bacteria that have gained resistance to penicillin, streptomycin, and other antibiotics.

## PLANT VIRUSES

Viruses that infect plants have been of interest to scientists primarily because of the extensive crop damage that is caused by these microorganisms. Worldwide the damage has been estimated to be $60 billion each year. Although the first plant virus, the tobacco mosaic virus, was identified more than 100 years ago, these viruses have not been characterized nearly as well as animal viruses.

Plant viruses are usually rod-shaped or filamentous, but the familiar icosahedron does occur, along with geminates (twin icosahedrons) and bacilliform. In contrast to bacteriophages, more than 75 percent of these viruses have RNA genomes. Consequently, these viruses usually have fewer than a dozen genes and are often extremely small, sometimes less than half the size of a typical bacteriophage.

Plants, unlike animals, have a cell wall that provides an effective barrier against viral infections. Viruses, carried to a plant by the wind or rain, cannot penetrate this outer defense. Consequently, plant viruses depend heavily on vectors to complete their life cycle. In most cases, insects such as aphids, white flies, and beetles transport the virus to its host and, by eating portions of the plant, open a path by which the viruses can enter the cells. Once inside a cell, viruses can travel throughout the plant by passing through the plasmadesma (small transport and communication pores) that link each cell, thus providing a channel through the otherwise impenetrable cell wall.

The plant that a virus infects depends almost entirely on the insect vector that it becomes associated with. That is, the virus can only infect those plants that its vector feeds upon. Virtually all plants fall prey to viral infections at some point in their lives.

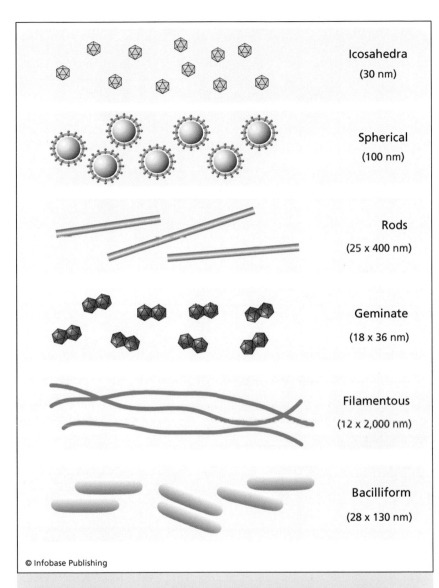

Icosahedra
(30 nm)

Spherical
(100 nm)

Rods
(25 x 400 nm)

Geminate
(18 x 36 nm)

Filamentous
(12 x 2,000 nm)

Bacilliform
(28 x 130 nm)

© Infobase Publishing

Plant viruses. Plant viruses are usually rod-shaped or filamentous, but the familiar icosahedron does occur, along with geminates (twin icosahedrons) and bacilliform. In contrast to bacteriophages, more than 75 percent of these viruses have RNA genomes. Consequently, these viruses usually have fewer than a dozen genes and are often extremely small, sometimes less than half the size of a typical bacteriophage.

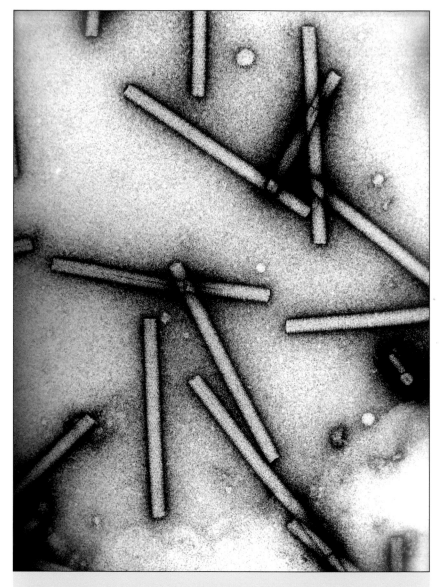

Tobacco mosaic virus (TMV). The image is a colorized transmission electron micrograph (TEM). Each virus consists of a protein coat with internal RNA genetic material. The effect of TMV infection varies from outright death of the host plant to severe lesions or mottling of the leaves. Plant viruses like TMV are transmitted by insects, pests, and nematode worms. Magnification: 300,000×. *(USGS)*

## CLASSIFICATION OF PLANT VIRUSES

| FAMILY | MORPHOLOGY | GENOME | HOSTS |
| --- | --- | --- | --- |
| Alphaflexiviridae | Filamentous | +ssRNA | Onions, mandarin oranges |
| Betaflexiviridae | Filamentous | +ssRNA | Apples, citrus fruits |
| Bromoviridae | Icosahedron | +ssRNA | Alfalfa, tomatoes, cucumbers, tobacco |
| Bunyaviridae | Spherical, envelope | -ssRNA | Tomatoes |
| Chrysoviridae | Icosahedron | dsRNA | Penicillium chrysogenum |
| Closteroviridae | Filamentous | +ssRNA | Grapes, beets, lettuce |
| Geminiviridae | Twinned icosahedra | ssDNA | Beans, beets, corn, tomatoes |
| Luteoviridae | Icosahedron | +ssRNA | Peas, barley, potatoes |
| Nanoviridae | Icosahedron | ssDNA | Bananas, clover |
| Potyviridae | Filamentous | +ssRNA | Blackberries, barley, potatoes, wheat |
| Virgaviridae | Rod-shaped | +ssRNA | Wheat, barley, peanuts, tobacco |

Note: All of the RNA genomes are linear, while the two DNA genomes are circular. There is also some variation regarding the exact form of the RNA genomes. That is, they may be single-stranded (ss), and if equivalent to an mRNA are called a plus (+) strand, or if they cannot serve as an mRNA they are called a minus (-) strand. Double-stranded (ds) RNA usually has both the plus and minus strands. Each host is infected by a different viral genus within the specified family.

## HUMAN VIRUSES

Human viruses come in many different shapes and sizes, from the familiar icosahedron to deadly serpentine giants that are nearly as large as the smallest cells. This variation, and the inherent complexity of these viruses, is due to the complexity of their eukaryote

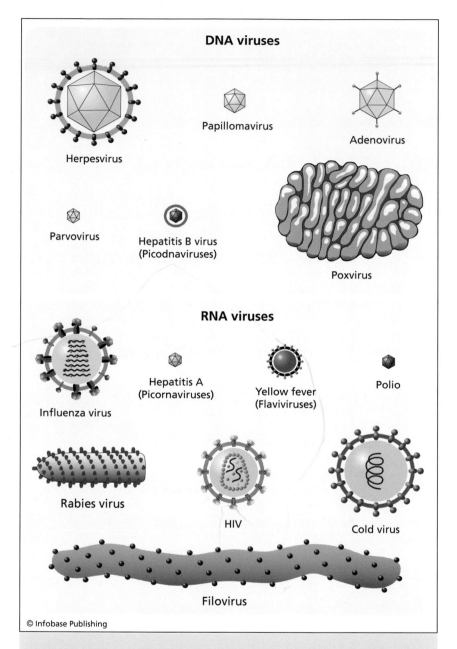

## DNA viruses

Herpesvirus

Papillomavirus

Adenovirus

Parvovirus

Hepatitis B virus
(Picodnaviruses)

Poxvirus

## RNA viruses

Influenza virus

Hepatitis A
(Picornaviruses)

Yellow fever
(Flaviviruses)

Polio

Rabies virus

HIV

Cold virus

Filovirus

© Infobase Publishing

Human viruses come in many different shapes and sizes, from the familiar icosahedron to serpentine giants that are nearly as large as small cells. This variation, and the inherent complexity of these viruses, is due to the complexity of their eukaryote hosts.

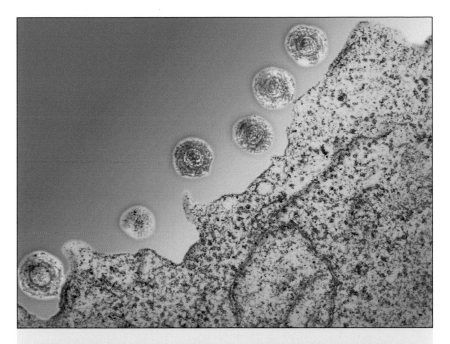

Human herpesvirus. The image is a color-enhanced transmission electron micrograph (TEM). This virus infects a variety of cells. The virus particles shown are budding from the surface of an infected lymphocyte. The "owl's eye" appearance of the virus particles is characteristic of the herpes family. *(Science Source/Photo Researchers, Inc.)*

hosts. Eukaryotes, being more complex than prokaryotes, are a more difficult host to invade. Consequently, throughout the long battle between animals and their viruses, each advance on one side has been countered by a new adaptation on the other. It is no coincidence that the largest and most complex viruses are those that infect humans, birds, and other animals.

The ICTV has classified 24 families of human viruses, 14 of which are discussed here (six families of DNA viruses and eight families of RNA viruses). These 14 families account for most of the human infectious diseases known to be of viral origin. The discussion begins with the DNA viruses.

## Herpesviridae

This is a large family of DNA viruses that have an enveloped ico-sahedral capsid, a linear single-stranded genome of about 200,000 bases (or nucleotides) that encodes more than 50 genes. The most important human herpesviruses include herpes simplex (oral and genital sores), varicella-zoster (chicken pox and shingles), Epstein-Barr virus (mononucleosis and neoplasms), and herpesvirus 8, which is associated with a cancer called Kaposi's sarcoma. These viruses have a relatively broad host range, being able to infect a wide variety of cell types, grow rapidly, and are very cytolytic. They also have the ability to establish lifelong infections and can undergo pe-riodic reactivation, particularly in immunosuppressed individuals (i.e., organ transplant patients and those suffering from HIV/AIDS).

Herpes viruses enter a cell by binding to a cell surface receptor called glycosaminoglycan and a co-receptor that is a member of the immunoglobulin superfamily. Binding to these receptors does not activate endocytosis. Instead, the bound virus fuses with the cell membrane, thus releasing the capsid into the cytoplasm, after which it moves into the cell nucleus. Transcription of viral DNA is car-ried out by cellular RNA polymerase II, but replication of the viral genome is performed primarily by viral enzymes. Viral enzymes also block cellular biosynthesis, so that the infected cell is eventu-ally killed. The infection cycle takes about 18 to 70 hours.

## Adenoviridae

These are medium-sized viruses that have a naked icosahedral cap-sid and a linear DNA genome consisting of 26,000 to 45,000 (bps). The capsid is about 90 nm in diameter and has spikes protruding from the 12 vertices. As indicated previously, these spikes possess the recognition sites for cellular receptors. The genome consists of 12 genes that code for several structural proteins and six enzymes. Adenoviruses infect mucous membranes in the respiratory tract, producing coldlike symptoms, and are also known to cause conjunctivitis (red eyes) and gastroenteritis. The infection cycle of this virus was discussed in chapter 2.

## Parvoviridae

Viruses in this family are the smallest of the DNA viruses. The capsid is icosahedral with a diameter of 18 to 26 nm. There is no envelope. The genome is single-stranded DNA consisting of 5,596 nucleotides that codes for only two proteins. These viruses are so simple they often require a helper virus before they are able to infect a cell.

The cellular target of these viruses is immature red blood cells, which they enter by endocytosis. Replication of this virus follows a path similar to that of the adenoviruses and consequently results in the death of the host cell. Thus, the resulting clinical symptom is chronic anemia, which is known to be fatal in the human fetus.

## Hepadnaviridae

The viruses in this family are relatively small (40 to 48 nm) with an enveloped icosahedral capsid. The double-stranded DNA genome is about 3,200 bp in size and encodes seven proteins, five of which are structural. These viruses cause acute liver infections, known as hepatitis. Persistent infections are associated with an increased risk of developing liver cancer. The members in this family are also known as the hepatitis B strain, to distinguish them from hepatitis A and C strains, both of which are RNA viruses. The infection cycle of the B strain is unusual in that the newly synthesized capsid buds from the Golgi complex, after which it acquires an envelope when it leaves the cell. The infection cycle of the A and C strains will be discussed with the RNA viruses.

## Papillomaviridae

This is a family of small viruses (55 nm) with a naked icosahedral capsid and a double-stranded genome that is 8,000 bp in size.

The genome encodes nine proteins, two of which form the capsid. Members of this family cause skin warts, laryngeal papillomas, and cervical cancer. Characterization of papilloma capsid proteins has made it possible for scientists to produce a vaccine, which destroys the virus, thus curing or preventing cervical cancer. The vaccine is called Gardasil and was approved for medical use in 2006. The infection cycle of this virus is similar to that of the adenoviruses.

## Poxviridae

These are among the largest and most complex of all viruses. The human smallpox virus is called variola (from the Latin "varius" meaning spotted or pimple). Viruses from this family also infect cows, monkeys, buffalo, sheep, and goats. The poxvirus is usually brick-shaped with a complex internal and external structure that does not conform to an icosahedral or helical form. At 230 nm × 400 nm, these viruses are almost as large as the smallest cells and can be seen as featureless particles under a light microscope. All other viruses require an electron microscope for visualization. Poxviruses have an envelope, covered by spiraling or irregular ridges, that encloses the genomic core.

The genome is linear ds DNA that consists of up to 375,000 bp and encodes about 100 genes. In addition to many structural proteins, the genome encodes a variety of enzymes that are involved in replicating the viral genome and transcribing mRNA. The genome of these viruses also encodes various growth factors that stimulate proliferation of infected cells, as well as proteins that inhibit the host's defense mechanisms.

Poxviruses infect the cell by fusing with the cell membrane or by activating endocytosis. The infection begins in the mucosa lining the upper respiratory tract. Soon after entry, the viral core is released into the cytoplasm. Viral-encoded enzymes dissolve the core, releasing the genome into the cytoplasm where the entire life cycle takes place. The RNA polymerase, stored in the core, produces mRNA that is translated by host ribosomes into viral proteins and

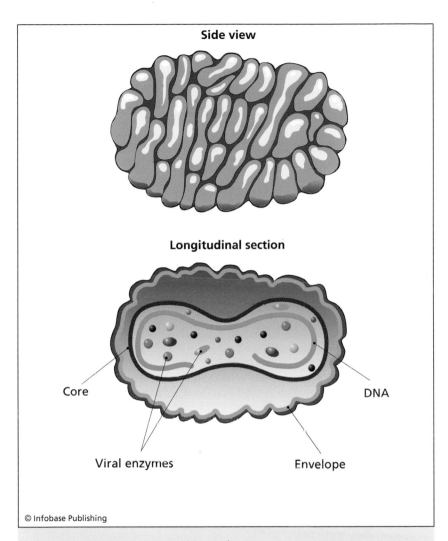

**Side view**

**Longitudinal section**

Core

DNA

Viral enzymes

Envelope

© Infobase Publishing

Poxvirus. These are among the largest and most complex of all viruses. The poxvirus is usually brick-shaped with a complex internal and external structure that does not conform to an icosahedral or helical form. The genome is linear double-stranded DNA that encodes a diverse group of enzymes that help it maintain the infection.

enzymes. One of these enzymes, DNA polymerase, replicates the viral genome in preparation for the formation of new viral particles, which leave the cell by budding. About 10,000 virions are produced

## CLASSIFICATION OF HUMAN VIRUSES

| FAMILY | MORPHOLOGY | GENOME | DISEASE |
|---|---|---|---|
| Herpesviridae | Icosahedron, envelope | DNA$^2$ | Herpes, chicken pox |
| Adenoviridae | Icosahedron, no envelope | DNA$^1$ | Colds |
| Parvoviridae | Icosahedron, no envelope | DNA$^2$ | Aplastic anemia |
| Hepadnaviridae | Icosahedron, envelope | DNA$^3$ | Hepatitis B |
| Papillomaviridae | Icosahedron, no envelope | DNA$^3$ | Warts, cancer |
| Poxviridae | Complex, envelope | DNA$^1$ | Smallpox |
| Orthomyxoviridae | Spherical, envelope | -ssRNA | Influenza |
| Picornaviridae | Icosahedron, no envelope | +ssRNA | Polio, hepatitis A |
| Flaviviridae | Spherical, envelope | +ssRNA | Yellow fever |
| Filoviridae | Helical, envelope | -ssRNA | Hemorrhagic fever |
| Coronaviridae | Spherical, envelope | +ssRNA | Common cold |
| Retroviridae | Complex, envelope | +ssRNA | HIV/AIDS |
| Paramyxoviridae | Pleomorphic, envelope | -ssRNA | Measles, mumps |
| Rhabdoviridae | Bacilliform, envelope | -ssRNA | Rabies |

Note: Human viruses have DNA or RNA genomes, but there is considerable variation concerning the exact form that they take. That is, the genome may be linear double-stranded (1), linear single-stranded (2), or circular double-stranded (3). The RNA genomes are all linear, but there is some variation regarding the exact form of the RNA. That is, the RNA may be single-stranded, and if equivalent to mRNA is called a plus (+) strand, or if it cannot serve as mRNA is called a minus (-) strand.

per cell, and more than half of these remain in the cell to produce additional virions. The initial infection spreads quickly to small blood vessels in the skin, mouth, and throat.

## Orthomyxoviridae

This is a family of influenza viruses, which marks the beginning of this chapter's discussion of human RNA viruses. Influenza is responsible for more than half of all acute infections each year in the United States. Worldwide, influenza has been responsible for millions of deaths and even with annual influenza vaccines still poses a major threat. There are three strains of these viruses called simply A, B, and C. Annual vaccines are directed at the virulent A strain, which is the topic of this section.

The flu virus is a spherical enveloped particle with an average size of 100 nm. The genome, contained in a helical nucleocapsid, consists of eight chromosomes of single-stranded negative-sense RNA, with an overall size of 13,600 nucleotides. The virus consists of nine different structural proteins, including the matrix protein, nucleocapsid proteins, and RNA polymerase. The envelope contains two glycoprotein spikes called hemagglutinin (HA) and neuraminidase (NA), both of which play an important role in the infection cycle. HA represents about 80 percent of the spikes and NA about 20 percent. HA is a trimer (i.e., three copies of an identical protein), whereas NA is a monomer. HA recognizes and binds to the cell-surface receptor, while NA appears to help the virus navigate through the mucin layer in the respiratory tract to reach the target cells. It also has a role in facilitating the exit of the virus from the cell. As discussed previously, there are many subtypes of these envelope proteins, and they serve as the source of the current naming convention for flu viruses. Currently, there are 15 known HA subtypes (H1 to H15) and nine NA subtypes (N1 to N9). These subtypes have been isolated from humans, birds, and other animals. Four HA subtypes (H1, H2, H3, and H5) and two NA subtypes (N1 and N2) are specific for humans. These subtypes generate a great number of variants, such as H1N1, H2N5, H5N1, and so on.

The segmented nature of the flu virus genome (i.e., the eight chromosomes) is the source of its variability and the reason novel flu vaccines are needed every year. When a cell is infected with two or more variants of a flu virus, the genomes can mix and reassort to produce a novel viral strain that may be more virulent than the original strains and at the same time resistant to current vaccines. The influenza virus enters a cell by receptor-mediated endocytosis, as discussed and illustrated in chapter 2.

## Picornaviridae

This family includes the dreaded poliovirus in addition to other viruses that cause meningitis, conjunctivitis, myocarditis, certain forms of hepatitis, and the common cold. Fortunately, most of the infections caused by these viruses are subclinical (i.e., do not cause an outward sign of disease). Indeed, many of these viruses are transient inhabitants of the human digestive tract and may be isolated from the nose and throat.

These viruses are relatively small (30 nm) bare icosahedrons (i.e., lacking an envelope) with a genome consisting of a single strand of positive-sense RNA. The genome ranges in size from 7,200 to 8,400 nucleotides and encodes seven proteins, four of which are essential for the construction of the capsid.

Picornaviruses bind to immunoglobulin receptors in the cell membrane that activate a conformational change in the capsid lead-

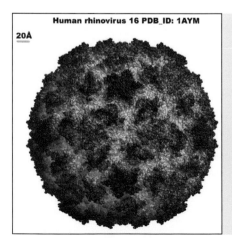

**Human rhinovirus 16 PDB_ID: 1AYM**

20Å

Rhinovirus. This virus, like the coronavirus, is responsible for the common cold. Rhinoviruses belong to the Picornaviridae family, which comprises small (about 30 nm in diameter) RNA viruses with a naked (no external envelope) capsid structure. Rhinoviruses are spread readily in air by droplets produced by coughing and sneezing. (Jean-Yves Sgro)

ing to the injection of the viral genome into the cytoplasm where the life cycle occurs. The RNA is translated by the host ribosomes into a large precursor protein that is processed to yield all of the structural and enzymatic viral proteins. One of the enzymes so produced is an RNA polymerase that replicates the viral genome. The viruses exit the cell by rupturing the membrane, thus killing the cell. The complete cycle takes about five to 10 hours.

## Flaviviridae

The viruses in this family are part of an ecological group, known as arboviruses, that are transmitted to humans by arthropods (ticks and mosquitoes) and rodents. This discussion is concerned with only two of these, the yellow fever and West Nile fever viruses, both of which are transmitted by mosquito vectors. These viruses have a spherical enveloped capsid that is 60 to 80 nm in diameter. The genome is positive-sense, single-stranded RNA consisting of 1,100 nucleotides that encodes nine proteins, three of which are located in the envelope.

Flaviviruses enter a cell by endocytosis. After the genome is released into the cytoplasm, it is translated by host ribosomes into a single large precursor protein that is cleaved by viral and host proteases to produce all of the viral proteins. The proteins destined for the envelope are sent through the endoplasmic reticulum for glycosylation. Replication of the genome occurs in the cytoplasm and the assembled virions leave the cell by exocytosis.

## Filoviridae

These viruses, like the flaviviruses discussed above, are transmitted to human hosts by mosquitoes or ticks. The two known members of this family, the Marburg and Ebola viruses, cause hemorrhagic fever that is nearly always fatal. They are easily transmitted and dangerous to work with.

Filoviruses (*filo*, Latin for worm) are pleomorphic but usually appear as large serpentine shapes that are about 80 nm wide and up to 1000 nm long. The large genome is single-stranded, negative-sense RNA 1,900 nucleotides in size that encodes seven proteins, one of which is located in the envelope. Very little is known about the life cycle of these viruses except that they leave the cell by exocytosis (budding).

## Coronaviridae

These viruses usually cause the common cold, but in 2003 a rare strain was responsible for a deadly outbreak of sever acute respiratory syndrome (SARS) in China, North America, and several other countries around the world. These are large enveloped viruses with a diameter of 120 to 160 nm. The single chromosome, the largest among the RNA viruses, is 3,200 nucleotides of single-stranded positive-sense RNA. Special viral proteins package the genome into a helical nucleocapsid. This virus consists of seven proteins, four of which are structural, including a glycoprotein that forms the envelope spikes.

Coronaviruses enter a cell by endocytosis, after which the capsid dissolves releasing the genome into the cytoplasm. Host ribosomes translate enough of the viral RNA to produce the viral RNA polymerase that synthesizes viral mRNA and effectively replicates the viral genome. The viral mRNA is translated by host ribosomes to produce the rest of the viral proteins. The spike proteins are sent through the Golgi complex for glycosylation (traffic through the ER and Golgi is illustrated in chapter 10). The life cycle of this virus is

*(opposite)* Coronavirus. The life cycle of this virus is unusual in that the nucleocapsid forms in the cytoplasm, but is then taken up by the Golgi. The viral envelope is thus obtained from the Golgi membrane in areas containing the viral spikes. The mature virions exit the Golgi in transport vesicles (bubbles) that carry them to the cell membrane. Fusion of these vesicles with the cell membrane releases the virions from the cell.

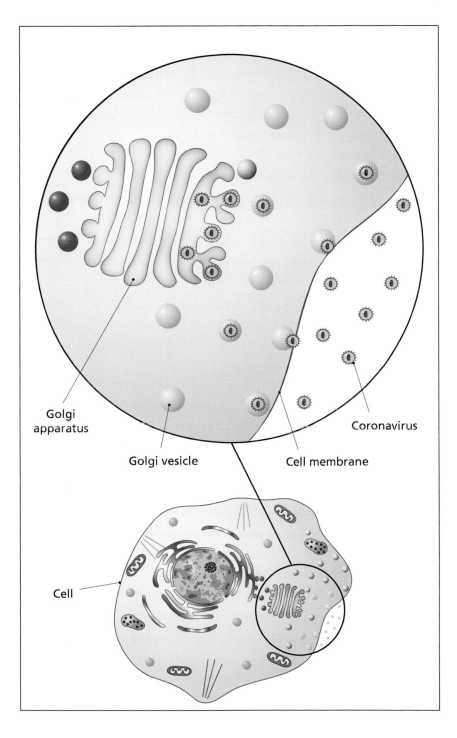

Golgi
apparatus

Golgi vesicle

Cell membrane

Coronavirus

Cell

unusual in that the nucleocapsid forms in the cytoplasm but is then taken up by the Golgi. The viral envelope is thus obtained from the Golgi membrane in areas containing the viral spikes. The mature virions exit the Golgi in transport vesicles (bubbles) that carry them to the cell membrane. Fusion of these vesicles with the cell membrane releases the virions from the cell.

## Retroviridae

Retroviruses, as discussed and illustrated in chapter 2, are a special group of RNA viruses that convert their RNA genome to DNA during the infection cycle. The DNA version of the viral genome is integrated into the host chromosome where it is transcribed to produce viral mRNAs. It also provides the virus with the opportunity of "hiding out" in the nucleus where it may lie dormant for many years. This stage in the life cycle of a retrovirus is one reason why HIV is such a difficult virus to eradicate. This will be discussed in chapter 7.

## Paramyxoviridae

These viruses are responsible for infections of the respiratory tract, known as measles and mumps, which were common in infants and children before the advent of modern vaccines. Structurally, these viruses are similar to flu and cold viruses. They have an enveloped spherical capsid that is about 150 nm in diameter and a negative-sense, single-stranded RNA genome consisting of 1,500 nucleotides. Most of these viruses have six to eight structural proteins, two of which form the envelope spikes.

Paramyxoviruses enter a cell by fusing with the membrane after which the nucleocapsid is released directly into the cytoplasm. The -ssRNA genome is converted to mRNA by a viral RNA polymerase stored in the nucleocapsid; it also replicates the genome. Thus, there is no need for the viral genome to enter the cell nucleus. Consequently, the entire life cycle occurs in the cytoplasm. The mRNA is translated by host ribosomes, and the envelope spike proteins are sent through the endoplasmic reticulum and Golgi complex for glycosylation. Once the nucleocapsids form, they acquire an envelope while exiting the cell as illustrated for the influenza virus (see figure on page 25).

## Rhabdoviridae

Rhabdoviruses are responsible for a fatal neurological disease known as rabies. Humans become infected with the rabies virus after being bitten by an infected animal, such as a dog, bat, or raccoon.

The rabies virus has an unusual bullet-shaped capsid, $75 \times 180$ nm, that is surrounded by an envelope containing a spiraling cluster of spikes. The genome is single-stranded negative-sense RNA of 1,200 nucleotides. The virus consists of five proteins, including an RNA-dependent RNA polymerase that is stored in the capsid.

The infection cycle begins with the virus binding to an acetylcholine receptor that is commonly found on neurons of the central nervous system. The bound virus is taken into the cell by endocytosis after which its genome, along with the attached RNA polymerase, is released into the cytoplasm. The RNA polymerase synthesizes viral mRNA and replicates the viral genome. The spike proteins are glycosylated in the Golgi complex and then deposited in the cytoplasm. Newly synthesized viral proteins and RNA genomes assemble nucleocapsids, which acquire envelopes while budding from the cell.

## SUMMARY

Viruses are extremely diverse. Nearly 10,000 different types of viruses have been identified, and these are believed to be only a fraction of the total population. Virtually every organism on Earth plays host to at least one type of virus. The major division in viral taxonomy is between prokaryote and eukaryote viruses. Bacterial viruses usually do not infect eukaryotes, and eukaryote viruses do not infect bacteria. Similarly, plant viruses are distinct from animal viruses, and within these major groups viruses are species specific. Animal viruses are the most complex, primarily because they are infecting eukaryote organisms that have a very effective immune system. The viral body, or capsid, not only protects the viral genome, but it also provides the equipment for entering the host cell. Once inside, the infection or life cycle involves commandeering cellular machinery to replicate daughter virions, which leave the cell by rupturing the membrane, thus killing the cell outright, or by exocytosis.

# A Brief History
# of Virology

Viruses have been with us for a very long time. Bacterial viruses probably appeared soon after the evolution of the first cells, more than 3 billion years ago. Animal viruses also have a deep history, possibly extending back 500,000 to 600,000 years. Human viruses are younger, but may have been around for the past 50,000 years. Viral infections from that period were likely mild infections, perhaps no more severe than a modern cold; most of them may even have been subclinical, meaning they did not produce observable symptoms.

Virulent human viruses probably did not emerge until the advent of agriculture, about 10,000 years ago. By that time, human populations had increased to the point that viruses could easily move from one host to another, thus increasing the rate at which they evolved. Human populations from earlier periods were small and spread out over an enormous area. Thus, even if a virulent strain appeared it would not have spread throughout the entire human population but would have been restricted to a small, possibly isolated, tribe.

This chapter will examine some of the evidence for viral infections in the ancient world and will discuss some of the highlights of microbiology and virology with an emphasis on smallpox, possibly the deadliest virus ever to appear on Earth. Despite the early development of a vaccine for this virus, which was the first vaccine ever produced, smallpox was killing 400,000 people every year before it was eradicated in 1979. With the invention of the microscope in the 17th century, scientists came to realize that they were surrounded by an invisible world full of strange creatures some of which, as they discovered, were responsible for diseases that had plagued humans for centuries.

## VIRAL DISEASES IN THE ANCIENT WORLD

The earliest evidence for human viral diseases comes from tombs in ancient Egypt. Ramses V, the fourth pharaoh of the Twentieth Dynasty, died in 1157 B.C.E. His mummified remains were discovered in 1898, and subsequent examinations showed that his face and neck were covered with pockmarks that bore a striking resemblance to smallpox. Subsequently, plagues believed to have been caused by smallpox were recorded at Syracuse in 595 B.C.E., Athens in 490 B.C.E., and throughout China in 48 C.E. The disease is believed to have originated in North Africa 6,000 to 7,000 years ago and to have spread with the expansion of the Moors from Saudi Arabia beginning in 647. By 731 it had reached Spain, and by the 16th century it was endemic throughout Europe and Asia.

Polio is another virulent viral disease that is known to have existed in the ancient world. This evidence also comes from Egyptian tombs, only in this case it is a carving on a wall rather than a mummy. An Egyptian stele dating from the Eighteenth Dynasty (1580 to 1350 B.C.E.) shows a priest with a withered leg, reminiscent of the effects that polio has on the limbs. Although it is possible that the deformity was caused by a rare birth defect or some other

Polio victim. Ancient Egyptian carving showing a priest with a shriveled leg typical of a recovered case of paralytic poliomyelitis (polio). Polio is an infectious disease caused by polioviruses that can permanently damage parts of the nervous system. In some cases, as seen here, it can cause paralysis. This bas-relief was carved in around 1500 B.C.E. *(SPL/ Photo Researchers, Inc.)*

trauma, it is very unlikely that the Egyptians would have taken the trouble to describe this condition if it were not a common malady.

Measles and chicken pox may have been common in the ancient world as well, but the evidence is circumstantial and rests on the fact that these diseases also produce sores on the skin similar to those caused by smallpox. Thus, some of the early descriptions of smallpox may have been outbreaks of these other viral diseases.

## SMALLPOX IN EUROPE AND THE NEW WORLD

By the 16th century, smallpox epidemics were common in Europe. The peak of these epidemics, which killed many thousands of Europeans, coincided with the exploration and colonization of the New World: North, Central, and South America. The introduction of this virus to the New World had a devastating effect on the Native populations.

In the early 1500s, Hernán Cortés, with fewer than 500 soldiers, defeated the great Aztec Empire, then being ruled by Montezuma. Cortés began his campaign by enlisting the aid of tribes long weary of Aztec rule, but in a major skirmish in Tenochitlan (now Mexico City), Cortés's troops suffered a great defeat, losing nearly one-third of its men, and were forced to retreat to a coastal settlement. Had the Aztecs pursued them, Cortés and all of his remaining force would almost certainly have been destroyed. The Aztecs had every intention of pursuing them and set about making the necessary arrangements. But before they could leave, smallpox, brought to their shores by the Spanish, hit the city. Most of the Aztec monarchy, along with half of the city's populace, perished from the disease. Aztec survivors described hellish scenes where the streets were filled with people dead or dying of the disease, with no one left to care for them. As one observer put it: "Dogs and vultures consumed the bodies." It has been estimated that 3 million Indians, about one-third of the total population, died from smallpox during this period.

By the 1700s, smallpox was the deadliest disease in the world, killing 400,000 people a year in Europe alone. It hit every level of society, with several monarchs, including Queen Mary of England and Louis XV of France, succumbing to the disease. In the new world, smallpox had reached Massachusetts as early as 1630. Puritan settlers described smallpox epidemics among the local tribes that killed everyone in the villages. Smallpox also exacted a heavy toll on the colonists and particularly the colonial army during the early days of the Revolutionary War. In some cases, half of the troops, including George Washington and members of Congress, came down with the disease. The only treatment in those days was self-inoculation with fluid from smallpox sores, which was believed to produce a milder form of the disease (in some cases it did, as explained in chapter 9). This procedure was known as variolation. As a consequence of the many fatalities, Washington ordered the entire Continental army variolated.

## THE FIRST VACCINE

In the 1790s, Edward Jenner, a British physician, noticed that dairy farm workers who had contracted cowpox, a relatively mild disease in humans, did not develop smallpox. In 1796, Jenner obtained the fluid from cowpox sores on one of his patients and transferred it to the skin of James Phipps, another of his patients. Jenner then variolated Phipps and found that he was immune to smallpox. Jenner tested this procedure on 19 other patients, all with the same results, and then published his results in 1798. Cowpox was known as *Variola vaccinae,* so the inoculum became known as a vaccine and the procedure, vaccination. Jenner was hailed as a hero and received letters of gratitude from European dignitaries, President Thomas Jefferson, and representatives of North American Indian tribes whose people had suffered so terribly from this disease. He also received many financial rewards and was appointed Physician Extraordinary to King George IV.

Despite the success of Jenner's vaccine, vaccination was surprisingly slow to catch on. During the U.S. Civil War, army regulations stipulated that all personnel were to be vaccinated, but this was not carried out with the necessary rigor and in many cases, owing to poor storage and a lack of refrigeration, the vaccine was spoiled or contaminated with other microbes. As a consequence, of the 600,000 Union and Confederate fatalities, two-thirds died of infectious diseases, and many of those were due to smallpox. Even President Lincoln was infected only days after delivering his famous Gettysburg Address and had to be quarantined in the White House for three weeks.

Although Jenner's vaccine was very effective, it would be almost 100 years before the fight against infectious diseases could be put on a firm scientific foundation. The building of that foundation began with the invention of the microscope.

## THE MICROSCOPE

No other instrument has had a greater impact on human lives and culture than the microscope. This instrument is at the heart of virtually every great discovery in the biological sciences since its invention in the 17th century. Prior to the microscopic age, the sum total of the world, aside from spiritual beliefs, consisted of what people could see with their eyes. Discoveries made with the telescope, another great optical instrument invented around the same time, cannot compare with the discoveries made with the microscope. The Moon, the Sun, the stars, and many of our neighboring planets were all there for us to see. The telescope simply gave us a closer look and helped us refine our understanding of those great celestial bodies. But the microscope revealed to us a world that no one had ever seen before, no one had even imagined it before: a world populated with the strangest creatures of such wild and diverse forms that there seemed to be no limit to their variability.

This is the world that Antoni van Leeuwenhoek, a humble Dutch lens grinder, discovered one evening in 1683. Leeuwenhoek's interest in the natural world was matched only by his dogged persever-

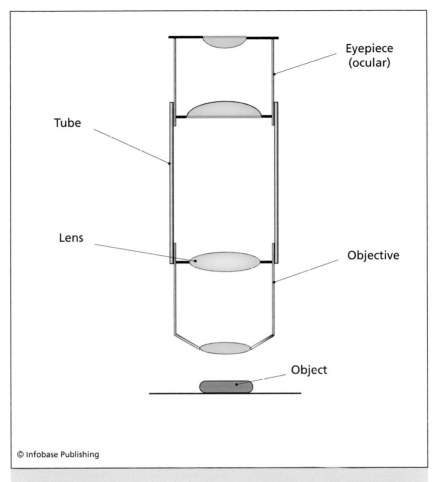

An 18th-century light microscope. The image is magnified by the objective and brought to focus at the eye by the eyepiece. The eyepiece and objective are compound lens systems that give high magnification but poor resolution and image quality.

ance. For many years he labored at his work bench, patiently melting fibers of glass to produce perfect little spheres that he used to make microscope lenses, capable of resolving an object 1,000 times smaller than a period on this page (about 1.0 μm). After polishing a lens, he mounted it between two copper or silver plates, stuck a

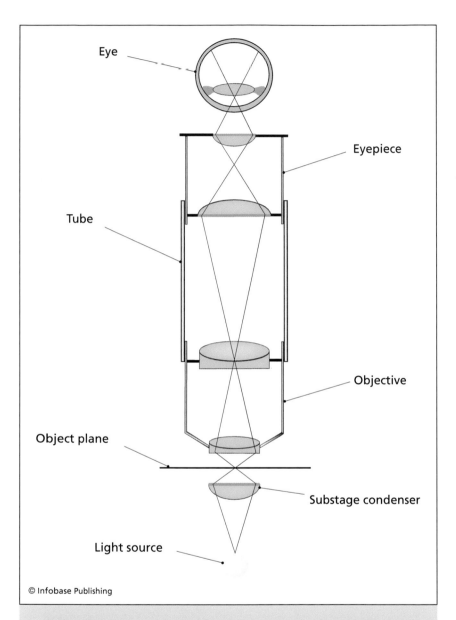

Eye

Eyepiece

Tube

Objective

Object plane

Substage condenser

Light source

A light microscope corrected for chromatic and spherical aberration. Light is focused on the object by a condenser to enhance image quality and contrast. Chromatic and spherical aberration are corrected at the objective with precisely ground lens-pairs or triplets (two or three lenses fitted together). High-power objectives (100×) may have up to 10 separate lenses to correct for aberrations. The example shown above is a 10× objective, consisting of four lenses.

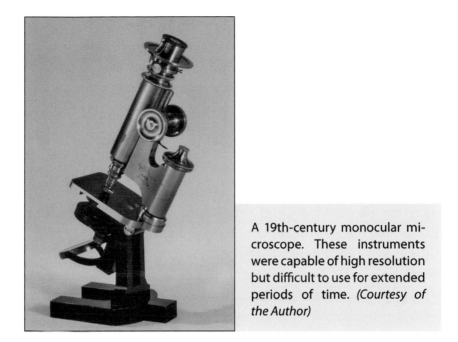

A 19th-century monocular microscope. These instruments were capable of high resolution but difficult to use for extended periods of time. *(Courtesy of the Author)*

sample holder onto one side, a handle on the bottom, and then held it in his hand or clamped it to his bench. One evening he looked at a sample of dental plaque taken from his own teeth, and the next morning he wrote an excited letter to the Royal Society of London describing the many "animalcules" that he had discovered.

Naturalists for the next 200 years tried to improve on Leeuwenhoek's microscope by using multiple lenses mounted inside a brass tube. These compound microscopes were more convenient to use, but the resolution was not as good as Leeuwenhoek's single-lens design. Adding two or more lenses together improved the resolving power theoretically, but the multiple lenses introduced optical artifacts that degraded the image substantially. In 1868, the German physicist Ernst Abbe discovered a way to construct objectives containing lens pairs and triplets that eliminated all of these artifacts. These lens systems, called apochromatic objectives, provide clear and undistorted images at the highest magnification possible (1,200×), while retaining a resolution of 0.4 μm; good enough for studying the microbial world.

## THE GERM THEORY

It was not long after Leeuwenhoek sent his letter to the Royal Society that physicians began to wonder if those "animalcules," now called microorganisms or microbes, could be responsible for human diseases. Indeed, the French physician Nicolas Andry de Bois-Regard suggested in 1700 that microbes, which he called "worms," were responsible for smallpox and other diseases. In 1847, the Hungarian physician Ignaz Semmelweis noticed that childbirth was relatively safe when handled by midwives but was often complicated by infections when handled in a hospital by doctors and medical students. Examining the situation closely, he found that infections occurred whenever the attending physicians had come directly from an autopsy. He concluded the infections were of microbial origin and insisted that the attending physicians wash their hands in water and lye before tending to their patients.

In the 1860s, the great French chemist Louis Pasteur showed that the growth of microorganisms was responsible for converting wine to vinegar and for spoiling beer and milk. His observations led to the now common practice of partially sterilizing milk and cheese (pasteurization). In 1864, he showed that a disease affecting the silkworm was caused by a microorganism. His results, generally accepted as the first, though not conclusive, statement of the germ theory, served as a warning to medical practitioners to take precautions against allowing microbes to enter the human body or to contaminate wounds. Soon after, physicians, led by the British surgeon Joseph Lister, began practicing antiseptic procedures during surgery.

In 1872, a German country doctor Robert Koch received a very fine microscope as a gift from his wife, which he used to study the relationship between disease and microorganisms. Following up on the germ theory speculations of others, Koch proved that anthrax, a fatal disease of cattle and humans, is caused by a bacterium, which he named *Bacillus anthracis*. Koch's germ theory postulates, published in 1890 and still in use today, are as follows:

1.  The microbe must be found in abundance in all organisms suffering from disease, but should not be found in healthy animals.
2.  The microbe must be isolated from a diseased organism and grown in pure culture.
3.  The cultured microorganism should cause disease when introduced into a healthy organism.
4.  The microorganism must be re-isolated from the inoculated, diseased experimental host and identified as being identical to the original specific causative agent.

Shortly after Koch's discovery, the Russian biologist Dimitri Ivanovski and the Dutch botanist Martinus Beijerinck showed that a disease affecting the tobacco plant was caused by an unusually small microorganism that could not be visualized under the microscope. To identify the microbe, a crushed extract from a diseased plant was passed through a special filter invented by Charles Chamberland, one of Pasteur's research technicians. This filter (now generally known as a millipore filter) had such fine pores that it could trap bacteria leaving the filtrate clear of anything larger than 0.5 μm. They found that the filtrate could be used to infect healthy plants. Initially, they thought it might have been a bacterial toxin in the filtrate that was responsible for the disease but eventually realized it was a new and extremely small microorganism. Beijerinck called the infectious filtrate a "contagious living fluid" which was later changed to "virus" after the Latin word for poison. The new microbe discovered by Ivanovski and Beijerinck is now known as the tobacco mosaic virus, a member of the Bromoviridae family.

Over the next 20 years, researchers used Chamberland filtration to identify many other viral diseases. But although the experiments with a filtrate satisfied the basic intent of Koch's postulates, it was not possible for virologists to satisfy them literally until after the invention of the electron microscope (EM) and special viral culturing techniques in the 1930s. The electron microscope, capable of resolving particles much smaller than bacteria, was invented by the German physicist Ernst Ruska and perfected by Reinhold Rudenberg, a German-American electrical engineer and the scientific director at the Siemens group of companies in Berlin. Rudenberg was especially interested in developing a microscope that could visualize viruses and in particular the poliovirus as it had infected and nearly killed his two-year-old son. His contribution to the development of the instrument was such that he obtained a patent for the instrument in 1938.

But even before the electron microscope was available, virologists used millipore filters to show that smallpox, polio, rabies, measles, influenza, and yellow fever were all caused by viruses and, by the late 1950s, had managed to produce vaccines for nearly all of them. By the late 1970s, researchers not only had excellent EM photographs to help them classify this new group of microbes, but DNA recombinant technology was making it possible to sequence the viral genomes and to characterize viral proteins. This not only improved the taxonomy but also made the search for an effective vaccine much easier. The problems involved in producing an effective viral vaccine will be discussed in chapter 9.

## FUTURE PROSPECTS

Deadly viral infections have been a plague on the human race for at least 5,000 years. The incidence and severity of viral epidemics increased dramatically as human population densities increased, thus making it easier for the viruses to move from host to host as well as increasing the rate at which they evolved. Only after the invention of the microscope and the discovery of the microbial world that it revealed was it possible for scientists to understand viral diseases and to find ways to combat them.

Today, virologists still face enormous challenges. Although smallpox was eradicated in 1979 and many of the viral diseases mentioned previously are being controlled with vaccines, deadly viruses, like HIV, keep emerging. After more than 20 years, a vaccine for this virus is still unavailable. More recent threats, such as the Ebola virus, the hantavirus, and the West Nile virus, are almost as deadly, but there is no vaccine available for any of them. And yet there is room for optimism. Methods for producing vaccines have improved considerably over the years, and ongoing viral research is clarifying the details of viral life cycles, which may lead to more effective ways to control or eradicate the deadliest of these microbes. The World Health Organization (WHO), although encountering some resistance from a few third world countries, is determined to eradicate polio, measles, and several other viral diseases. With worldwide vaccination programs, the problem is more a matter of trust (that the vaccine will do what it is supposed to do without causing any harm) than it is a matter of financing the effort and coordinating the delivery. The reluctance of some people and governments to use available vaccines has been the root of the problem since Jenner's day, but with education and gentle persistence the effort will succeed in the end.

# Viruses in the Sea

Viruses, like cells before them, came from the sea, and it is in that environment that scientists hope to find examples of truly ancient virus-host interactions. Marine viruses infect bacteria, archaea, and eukaryotes and are important members of the marine food web. Traditionally, marine virology has focused on viruses as pathogens of aquatic organisms, but recent studies have shown that the interactions between viruses and their hosts are much more complicated.

Marine viruses are more prevalent than their hosts, and they may have accelerated the evolution of prokaryotes through horizontal gene transfer, a process that is known as genetic transduction and transformation. That is, a virus may acquire a gene while infecting one bacterium and then transfer it to another cell (transduction), where it may have transformed the new host, giving it a selective advantage. Viruses certainly do parasitize marine organisms, and many of those organisms are killed in the process. But collectively, viruses play an important role in the maintenance of the marine environment and as genetic reservoirs.

## VIROMES AND MICROBIOMES

Recent studies have attempted to estimate the extent of viral diversity in the marine environment. This has been difficult because viruses do not have a universally conserved gene, like the ribosomal DNA genes in cellular organisms, which could be used to estimate diversity. As an alternative, researchers have isolated entire viral communities from which shotgun sequence data are obtained. In this procedure, the genomes in the pooled community are sequenced at random, and then with the help of computers the

Diatoms. These organisms are part of the phytoplankton, a diverse group of marine protozoans that are capable of photosynthesis. Phytoplankton, together with heterotrophic protozoans and bacteria, constitute the marine microbiome. Magnification is approximately 500×. *(Dr. Neil Sullivan/NOAA, Department of Commerce)*

individual genomes are reconstructed, characterized, and enumerated. The resulting data is called a marine viral metagenome or virome. A similar analysis performed on bacteria or eukaryotes is called a microbiome. Analysis of viromes has shown that marine viruses are extremely diverse with more than 5,000 viral genotypes in 26 gallons (100 l) of seawater and up to 1 million species in 2.2 pounds (1 kg) of marine sediment. Viromes collected from around the globe have shown that the various species are everywhere, but that there are some variations with regards to relative abundance.

## VIRUS-MEDIATED GENE TRANSFER

Viruses are known to carry and transfer a variety of host genes. It is generally assumed that gene transfer of this sort is beneficial for the virus but not for the host. Recently, however, studies have shown that in some cases transformation, involving the horizontal transfer of genes from a virus to its host, can actually improve the host's metabolic vigor and extend its environmental range.

Two examples are the photosynthetic cyanobacteria *Synechococcus* and *Prochlorococcus,* which together account for almost 25 percent of global photosynthesis. Sequence analysis of the viruses that infect these cells (called cyanophage) has shown that the viruses often carry some or all of the genes needed for photosynthesis. These genes include the high-light-inducible gene *(Hli)* and two genes that encode the photosystem II (PSII) proteins D1 and D2 *(PsbA* and *PsbD).* It has been determined that about 60 percent of the photosynthesis genes in the marine environment are actually from cyanophage and that about 10 percent of total global photosynthesis is performed by *PsbA* originally from cyanophage. Since these phages carry photosynthesis genes, it is likely that heterotrophic marine bacteria (i.e., cannot photosynthesize) have been converted to autotrophs (i.e., can photosynthesize) after being infected by these viruses. Indeed, many studies have shown an exchange of phage PSII genes between bacterial hosts. This transformation broadens the ecological niche of the bacterium, since it is no longer restricted to areas high in dissolved nutrients but can spread out over the surface of the ocean. It also suggests that photosynthesis, once it evolved in certain prokaryotes, may have spread through the ecosystem much faster than would have been possible otherwise. Thanks to the cyanophage, the greening of Earth, or at least the marine environment, may have been a very rapid process.

Remarkably, viruses have also transferred photosynthesis genes to metazoans. The best-studied example is the sea slug, *Elysia chlorotica.* These animals eat algae and then harvest the chloroplasts through specialized epithelial cells in the gut. Once internalized (through phagocytosis), the chloroplasts are maintained within the animal's cells for many months during which time the slug obtains energy directly from the Sun, as though it were a plant. But scientists have known for some time that chloroplasts encode only some of the photosynthesis proteins, with the rest being encoded by the cell nucleus. On careful examination, researchers have found that the slug genome contains the rest of the needed genes and that these genes were obtained from the algal nuclear genome by horizontal transfer, presumably mediated by a eukaryotic virus. The virus has been observed in tissue samples taken from the slug but has not been identified.

Viruses are also known to have transferred genes to humans. Analysis of the human genome has shown that 220 of our genes were obtained by horizontal transfer from bacteria, rather than ancestral or vertical inheritance. In other words, humans obtained these genes directly from bacteria. Scientists know this to be the case because while these genes occur in bacteria they are not present in yeast, fruit flies, or any other eukaryotes that have been tested. Researchers believe that viruses collected these genes from their bacterial hosts and then transferred them to human cells in a kind of natural gene therapy. This was likely accomplished in a stepwise procedure, whereby bacteriophage picked up the genes from their hosts and transferred them to protozoans, after which eukaryote viruses picked them up from the protozoans and transferred them to humans and possibly other animals.

Transformation is also known to convert a benign microorganism to a pathogen. The best-known marine example is the bacterium *Vibrio cholerae,* a harmless near-shore microbe that is converted to a deadly pathogen by incorporating phage cholera toxin genes. When people drink the water containing these transformed bacteria, the toxin, which is known to be a protein complex, is absorbed by cells in the intestine. Once inside the intestinal cells, the toxin activates a cellular pump that secretes water and salts into the lumen of the gut, resulting in a massive and often fatal diarrhea. Virome analysis has shown that viruses contain many other bacterial virulence genes, including some that confer antibiotic resistance and toxicity, as well as enhancing the ability of bacteria to invade potential hosts. Bacteria that acquire these genes extend their ecological niches, but that is usually bad news for humans and other animals infected by these transformed pathogens.

## MARINE VIRUSES ARE NOVEL GENE BANKS

Viral genomes contain many host genes that are transferred to bacteria, protozoans, and metazoans. Recent estimates show that as many as $10^{25}$ genes are moved by horizontal transfer from virus to host each year in the world's oceans; a truly colossal undertaking. To put this into perspective, the human genome contains about $3 \times 10^4$ (30,000) genes, so every year marine viruses are transducing the equivalent of about $10^{21}$, or a billion trillion, human genomes. A recent study of deep-sea viruses showed that about 75 percent of sequences from the virome were novel and could not be identified in the GenBank database (a DNA database maintained by the National Institutes of Health). This abundance of novel sequences suggests that deep-sea viral communities could be a store of genes that the local bacterial population needs in order to adapt to the high pressures, high temperatures, and high concentrations of sulphides, iron, and salt that are found at great depths. Viruses serving as repositories of novel genes are known to exist in surface ocean communities as well.

Not all of the transduced genes are being moved by viruses. Scientists have identified a new viruslike particle known as a generalized transducing agent (GTA). GTAs are similar to phage but are smaller with a bare-capsid diameter of 30 to 50 nm. Their DNA genome is also small, consisting of about 4,000 nucleotides. The unusual feature of these particles is that they only carry host genes and thus appear to be at a transitional stage, where they are more than a plasmid but have not yet become viruses. GTAs were originally identified in the marine bacterium *Rhodobacter capsulatus* but are now known to be widespread among bacteria and archaea. They are believed to be very ancient, likely originating with the ancestral prokaryotes. Studies have shown that genes encoding GTAs are found in most bacterioplankton.

## VIRUSES AS MARINE PREDATORS

Terrestrial ecosystems derive much of their stability from predator-prey relationships. Were it not for predators, herbivore populations would increase to such an extent that the vegetation upon which they feed would all be destroyed by overgrazing. This happened in the early part of the 20th century when the loss of large predators (wolves, bobcats, mountain lions) in Appalachian states allowed the white-tailed deer population to increase to such an extent that it came close to destroying large portions of its habitat. Terrestrial predators also maintain a species balance that is crucial for a healthy ecosystem. This was illustrated by the loss of the gray wolf in Yellowstone National Park in the 1920s and the beneficial changes that occurred after the wolf was reintroduced to that ecosystem in 1995.

In the ocean, it was always assumed that bacterial populations were being maintained by predatory protozoans, and that the protozoan populations were being controlled by zooplankton. But research has shown that the predators managing both of these populations are viruses. In the surface layers, it has been estimated that viruses kill about 40 percent of the marine bacteria and eukaryotic plankton on a daily basis. Aside from the fact that this helps control the size of these populations, it also serves a very important role in the recycling of nutrients and energy. By the lysis of cells near the surface, viruses help reduce the rate at which carbon compounds sink to a deep ocean layer where they may be trapped for hundreds of years. In some cases, released nutrients such as iron satisfy the requirements of other organisms. Thus, marine viruses play a crucial role in providing nutrients that may be used by the rest of the microbial community.

## MARINE VIRUSES AND DISEASE

Viruses infect everything in the ocean from bacteria to whales. Knowledge of the diseases they produce comes almost entirely from the fishing industry and normally involves obvious signs of infections or a large number of fatalities. But very little is known about the viruses involved, their mode of infection, or their reservoirs. Some of these viruses may pose a risk for humans. The distemper viruses, for example, are believed to cycle between marine and terrestrial mammals. Similarly, it is known that marine birds carry the avian flu virus, particularly the H5N1 strain that has been involved in human flu pandemics. Whether the ocean environment will be the source of a future human epidemic is a potential problem that needs further investigation.

## SUMMARY

An understanding of marine viruses has improved considerably over the past 10 years. Viruses are not only abundant and active in the sea, but it is now clear that they possess the greatest genetic diversity in the ocean, are effective predators that regulate planktonic populations, and play an important role in maintaining global geochemical cycles. In addition, through genetic transduction, they have altered the adaptability of many organisms and have almost certainly accelerated the rate at which individual organisms and the biosphere as a whole has evolved over the past 2 billion years.

# Viruses in
# Biomedical Research

Viruses are not just about sickness and disease. Viruses and their transposon ancestors forced plants and animals to develop genomes that could accommodate wandering genetic elements. Coevolution of these organisms created plant and animal genomes that contain safe insertion sites where a virus can land without damaging the host's cellular genes. This has proved to be an extremely important feature, for scientists are now able to use viruses to produce stem cells, to cure genetic diseases, and to create gene libraries that have revolutionized biomedical research.

## BIOTECHNOLOGY

The primary function of biotechnology, which includes recombinant DNA technology, is the study of the cell and its genes. The first step in the procedure is to isolate the gene of interest, after which it is amplified (to increase its amount) and sequenced. Once this has

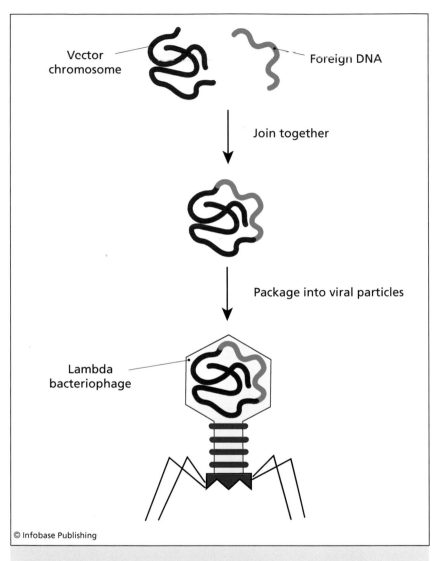

DNA cloning. The isolated phage genome is cut with a special enzyme and added to the purified cellular DNA. Under the right conditions, a single piece of cellular DNA will be incorporated into each of the viral chromosomes. The DNA so produced is called recombinant DNA because it contains both viral and cellular genes. The recombinant DNA is then added to a special packaging mixture to produce whole virus particles, which are used to infect bacteria.

been accomplished, pieces of a specific gene can be used as a probe to analyze its expression pattern in a given cell, tissue, or even an entire organism. Viruses play an important role in this technology for without them the first step would be almost impossible.

Isolating a specific gene begins with the isolation of total DNA from cells or tissues of interest. The DNA is cut up into small pieces (about 10,000 base pairs [bp]) with special enzymes. Next, DNA is isolated from a suitable virus, usually a bacteriophage called lambda. The lambda genome is double-stranded linear DNA of about 40,000 bp, much of which can be replaced with foreign DNA without sacrificing the ability of the virus to infect its host, the bacterium *E. Coli*. The isolated phage genome is cut with a special enzyme and added to the purified cellular DNA. Under the right conditions, a single piece of cellular DNA will be incorporated into each of the viral chromosomes. The DNA so produced is called recombinant DNA because it contains both viral and cellular genes. The recombinant DNA is then added to a special packaging mixture to produce whole virus particles, which are used to infect bacteria.

The number of phage particles increases tremendously as the infection cycle progresses, and as the population grows so does the number of copies of the foreign DNA. Each time the phage replicates its genome, it replicates the piece of foreign DNA as well. This process is called DNA cloning because the subsequent copies of the foreign DNA are identical to the original. The expanded population of phage, each with a different piece of foreign DNA, represents a library of the cell's entire genome from which other genes or gene fragments may be isolated for any number of studies that could improve our understanding of cell biology.

Prior to the invention of biotechnology, biologists had only a basic understanding of the cell: They knew the DNA was located in the nucleus, that the cell was surrounded by what appeared to be a featureless membrane, and that the cell interior was full of structures called organelles, but their functions were largely unknown.

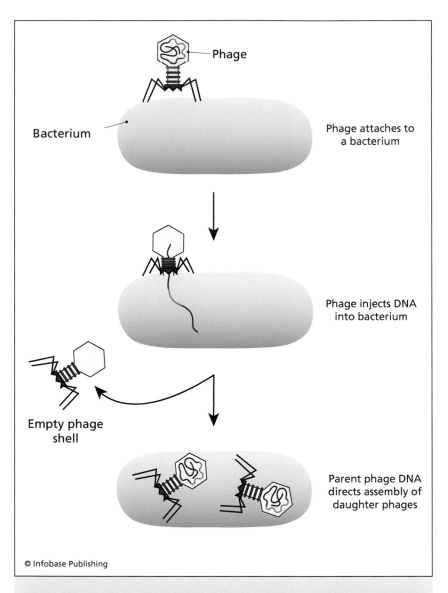

Phage

Bacterium

Phage attaches to
a bacterium

Phage injects DNA
into bacterium

Empty phage
shell

Parent phage DNA
directs assembly of
daughter phages

© Infobase Publishing

DNA library. Recombinant bacteriophages are added to a solution containing host bacteria. The number of phage particles increases tremendously as the infection cycle progresses, and as the population grows so does the number of copies of the foreign DNA. The expanded population of phage, each with a different piece of foreign DNA, represents a library of the cell's entire genome.

Today, thanks to biotechnology, scientists have sequenced the entire human genome, as well as the genomes of many other organisms. They have determined the function of virtually every cellular organelle, and they have shown that the cell membrane, far from being featureless, contains thousands of glycoproteins that give the cell its eyes, its ears, and the equipment it needs to capture food and to communicate with other cells.

By studying the cell, scientists improve our understanding of the living world and, in particular, our understanding of plant and animal physiology, genetics, and biochemistry. This wealth of information has revolutionized the biological and medical sciences. For human society, this knowledge translates into a dramatic reduction in mortalities due to infectious diseases and medical disorders. Thus, it is ironic that biotechnology and the insights and medical therapies it has provided ultimately depend on a parasitic infectious microorganism.

## GENE THERAPY

An illness is often due to invading microbes that destroy or damage cells and organs in our body. Cholera, smallpox, measles, diphtheria, AIDS, and the common cold are all examples of infectious diseases. Such diseases may be treated with drugs that will in some cases remove microbes from the body, thus curing the disease. Unfortunately, many diseases are not of the infectious kind. In such cases there are no microbes to fight, no drugs to apply. Instead, physicians are faced with a far more difficult problem, for this type of disease is an ailment that damages a gene. Gene therapy attempts to cure these diseases by replacing, or supplementing, the damaged gene.

When a gene is damaged, it usually is caused by a point mutation, a change that affects a single nucleotide. Sickle-cell anemia, a disease affecting red blood cells, was the first genetic disorder of this kind to be described. The mutation occurs in a gene that codes for the β (beta) chain of hemoglobin, converting the codon

GAG to GTG, which substitutes the amino acid valine at position 6, for glutamic acid. This single amino-acid substitution is enough to cripple the hemoglobin molecule, making it impossible for it to carry enough oxygen to meet the demands of a normal adult. Scientists have identified several thousand genetic disorders that are known to be responsible for diseases such as breast cancer, colon cancer, hemophilia, and two neurological disorders, Alzheimer's disease and Parkinson's disease.

Given their talents for entering cells, viruses are ideal candidates for gene delivery vehicles (also known as gene vectors). But there are two major problems to overcome before they can be used safely: First, the ability of the virus to replicate its own genome must be blocked, along with the production of viral mRNA that codes for proteins that maintain the infection and help the virus escape from the cell. Second, the therapeutic gene has to be inserted into the viral genome in such a way that it will not inhibit the formation of a normal capsid, since this is the part of the virus that is essential for cell entry.

Adenovirus type 2 (AD-2) and a retrovirus called Murine (mouse) leukemia virus (MLV) are used in more than 90 percent of all gene therapy trials to date. Production of viral gene vehicles is carried out in a test tube. Viral genes needed for replication and the maintenance of infection are removed, after which the therapeutic gene is inserted into the viral chromosome. The hybrid chromosome is added to a test tube and mixed with purified viral capsid proteins, leading to the auto-assembly of viral particles. If this is done properly, the virus will be able to enter the cell to deliver the gene, but it will not harm the cell nor will it be able to reproduce itself. The recombinant viruses, or gene vectors, may be introduced into cultured cells suffering from a genetic defect and then returned to the patient from whom they were derived (ex vivo delivery). Alternatively, the vector may be injected directly into the patient's circulatory system (in vivo delivery). The ex vivo procedure is used

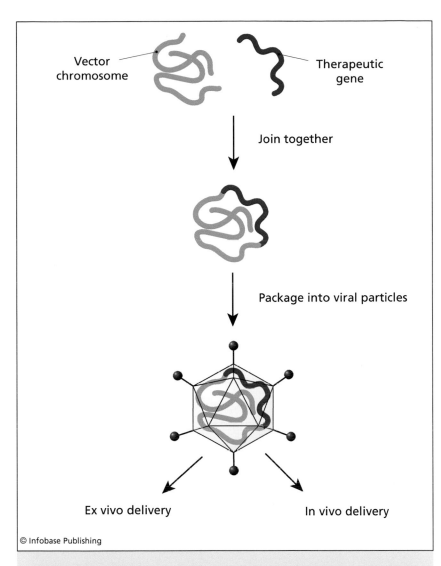

Vector preparation and delivery. A viral chromosome and a therapeutic gene are cut with the same restriction enzyme and the two are joined together, after which, the recombinant chromosome is packaged into viral particles to form the vector. The vector may be introduced into cultured cells, and then returned to the patient from whom they were derived (ex vivo delivery), or the vector may be injected directly into the patient's circulatory system (in vivo delivery).

when the genetic defect appears in white blood cells or stem cells that may be harvested from the patient and grown in culture. The in vivo procedure is used when the genetic defect appears in an organ, such as the liver, brain, or pancreas. This is the most common form of gene therapy, but it is also potentially hazardous because the vector, being free in the circulatory system, may infect a wide range of cells, thus activating an immune response that could lead to widespread tissue and organ damage.

AD-2, although naturally adapted to infecting the upper respiratory tract, has been used in trials that targeted T lymphocytes, liver, skin, and a variety of tumor cells. An important consideration when using this virus is the amount to give the patient. In a trial attempting to treat a liver ailment, for example, the recombinant AD-2 is injected directly into that organ. If the number of viral particles injected is correct, the liver receptors will bind up all of the viral particles. If the amount is too low, too few cells will take up the virus, so expression of the therapeutic gene will be insufficient to treat, or cure, the disease. If the amount is too high, viral particles will spill out into the general circulation and infect a variety of cells. Being crippled, these viruses cannot damage the cells they infect, but their presence can lead to a potentially deadly immune response as T cells detect and destroy infected cells. In extreme cases, this can lead to the destruction of entire organs and the death of the patient.

While the adenovirus has proved to be a good delivery vehicle, the expression of the therapeutic gene tends to decline after a week or two. This is believed to be due to the extra chromosomal life cycle of this virus. That is, the viral chromosome enters the cell nucleus, but it does not integrate into a host chromosome. Under these conditions, the cell's machinery does not continue transcribing the therapeutic gene. Moreover, AD vectors are inefficient at infecting some cells, and they tend to activate an antivector immune response.

Consequently, most clinical trials use a retrovirus as a delivery vehicle. Indeed, the retrovirus is not only the current favorite but was the first to be developed for gene therapy. These viruses are very efficient at infecting cells of the immune system, and they do not elicit as strong an immune response as do other vectors. Moreover, the retroviral life cycle includes integration of its genome into the host chromosome. Once it is in the chromosome, the therapeutic gene is expressed at a steady rate.

Unfortunately, in many cases, the rate at which a therapeutic gene is expressed by a retroviral vector is too low to cure the patient or even to alleviate some of the symptoms. In addition, there is always some apprehension about using an integrating virus, because if something goes wrong there is at present no way to get it out again. This is particularly worrisome, since in an attempt to increase expression of the therapeutic gene, some gene therapy trials use retroviral vectors that are replication competent (can still reproduce).

The justification for designing replication competent retroviral vectors is that these viruses do not kill the cell when they exit. If its pathology-inducing genes are removed, reproduction of the vector and its movement from cell to cell are of no concern. Vector reproduction leads to an increased number of cells being infected and thus increases the amount of therapeutic protein being synthesized with subsequent benefits for the patient. There is, however, the possibility that one of these vectors will encounter another virus infecting the patient and, through genetic recombination, become pathogenic and possibly deadly.

An alternative approach involves genetic engineering of hybrid retroviruses that might produce large quantities of the therapeutic protein while being unable to replicate. To this end, scientists have created an Ebola-HIV viral hybrid to be used as a novel gene delivery vehicle. Scientists know that both viruses are deadly and exceptionally talented when it comes to infecting cells. The hybrid appears to work well in animal experimentation, but whether it will ever be approved for use in human gene therapy trials is another question.

The first gene therapy trial, conducted in 1990, used ex vivo delivery. This trial cured a young patient named Ashi DeSilva of an adenosine deaminase deficiency that affects white blood cells. Many other trials since then have either been ineffective or were devastating failures. Such a case occurred in 1999, when Jesse Gelsinger, an 18-year-old patient suffering from a liver disease, died while participating in a gene therapy trial. His death was caused by multiorgan failure brought on by the viral vector. In 2002, two children being treated for another form of immune deficiency developed vector-induced leukemia (cancer of the white blood cells). Subsequent studies, concluded in 2009, appear to have resolved these problems. Gene therapy holds great promise as a medical therapy. In the United States alone, there are currently more than 1,000 trials in progress to treat a variety of genetic disorders.

## STEM CELL THERAPY

Stem cells are special multitalented cells that could be used to treat many medical disorders. Unfortunately, they are also very controversial. A casual observer may wonder how this could be. After all, stem cell therapy holds the potential for curing terrible diseases such as cancer, Parkinson's disease, and Alzheimer's disease. In addition, this therapy may offer hope to those who have lost the use of their arms or legs because of a spinal cord injury, and someday it may be used to reverse the aging process. What could be wrong with such a promising procedure?

The controversy, and much of the confusion, centers around the fact that there are three different kinds of stem cells: adult stem cells (AS cells), embryonic stem cells (ES cells), and manufactured stem cells. The adult stem cells are isolated from adult tissues, such as bone marrow, and from umbilical cord blood. The embryonic stem cells are isolated from two- to three-day-old human embryos. The embryos do not survive the harvesting of the stem cells and many people believe that it is unethical and immoral to kill a human embryo for its cells. Manufactured stem cells, the newest member of the stem cell family, are produced in the laboratory by

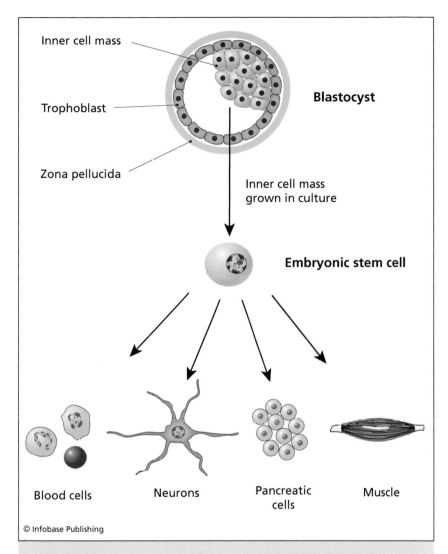

Inner cell mass

Trophoblast

Zona pellucida

**Blastocyst**

Inner cell mass
grown in culture

**Embryonic stem cell**

Blood cells          Neurons          Pancreatic
cells          Muscle

© Infobase Publishing

Differentiation of embryonic stem cells. Embryonic stem cells are obtained from the inner cell mass of a blastocyst. When cultured, these
cells can differentiate into many different kinds of cells, representing
the three germ layers.

reprogramming skin cells. The use of embryonic stem cells is very
controversial, whereas the use of adult or manufactured stem cells
is not.

Many scientists believe that ES cells offer the best hope for effective medical therapies, because these cells possess a high degree of developmental plasticity (i.e., can produce a wide variety of cell types) and, when injected into experimental animals, sometimes repair, or try to repair, whatever damage is present. The results of some of these experiments have been so encouraging that scientists now call ES cells the "gold standard" of stem cell research.

But there is a darker side to this story. The elasticity of an ES cell is in fact its Achilles' heel. Scientists may inject these cells into a mouse or a human, but they cannot control them. Studies in mice have shown that while injected ES cells initiate some repairs, they are also busy forming cancerous tumors, called teratomas, in the brain and in other parts of the body. A 17-year-old Israeli patient was injected with ES cells in 2001 to treat a neurological disorder. In 2005, after complaining about chronic headaches, he was examined by physicians at Sheba Medical Center in Tel Aviv who found tumors in his brain and spinal cord. Subsequent tests traced those tumors to the transplanted ES cells.

Fear of teratomas is the main reason why the FDA has been so reluctant to approve clinical trials involving ES cells. Of the more than 2,000 American clinical trials involving stem cells, only one is using ES cells. This one exception, pioneered by the biotech company Geron Corporation, gained approval only because the ES cells were coaxed into producing neuron precursor cells before being used, thus reducing the risk of teratoma formation. The FDA also insisted that the cultures of neuron precursors be carefully screened to ensure they are free of undifferentiated ES cells. It has been suggested that a way around this problem is simply to use AS cells as a source for these therapies. These stem cells are already being used to treat leukemia, and recent studies suggest they will soon be used to treat spinal cord injuries, multiple sclerosis, and kidney disease. But some researchers argue that AS cells are not suitable for the production of neural precursors that are needed to treat Alzheimer's or Parkinson's disease.

Thus, it came as a relief when in 2007 scientists in Japan produced embryo-like stem cells from ordinary skin cells. The official name for these cells is induced pluripotent stem cells (iPS cells). "Induced" because scientists engineer the reprogramming that converts the skin cells to stem cells, and "pluripotent" because they have a high degree of plasticity. Indeed, iPS cells are just as versatile as ES cells. Moreover, they are easy to produce, grow well in culture, and are free of ethical controversy because they are not obtained from human embryos.

These iPS cells are produced using a virus to introduce four transcription factors into skin cells, which alter the gene expression of the skin cells in such a way that they are reprogrammed to become undifferentiated (i.e., embryo-like) stem cells. Transcription factors are gene-regulatory proteins and were carefully chosen to ensure that the skin cells were converted into normal cells, rather than cancer cells or some other abnormal cell type. Now, iPS cells have been used to generate a wide range of cell types, including neurons, and are expected to revolutionize the field of stem cell research.

## SUMMARY

Viruses have played a crucial role in the development of some of the most important technological achievements of the past 50 years. Biotechnology, providing a way to clone, sequence, and characterize genes and entire genomes, has revolutionized biomedical research. Gene therapy, providing ways to correct or supplement mutated genes, has almost unlimited potential. This procedure has already been used to treat a variety of inherited diseases and most recently was used to restore vision in patients suffering from partial degeneration of the retina. Stem cell therapy has a potential equal to if not surpassing that of gene therapy. Although AS cells have been used to treat a variety of ailments, real excitement entered the field with the viral-assisted production of iPS cells. With these cells, a cure for many of the disorders mentioned above, such as Alzheimer's disease or a damaged spinal cord, may finally be at hand.

# Viral Diseases

Viral diseases have killed more people over the centuries than wars and all other diseases combined. The outstanding feature of a viral disease is that the infectious agent attacks at the heart of a cell. Unlike a bacterial or fungal infection, a virus can commandeer and subvert the cellular machinery for its own gain. In some cases, it can even bury its genome within the cell nucleus where it can hide from the human immune system until it is ready to launch a new attack. Fighting such an adversary has proved to be an immense challenge. The list of viral diseases is a very long one: smallpox, polio, rabies, influenza, AIDS, and yellow fever are but a few examples. Currently, at least 65 human diseases are known to be caused by viruses. This chapter describes 14 of the most prominent viral diseases. A few of these infections, such as the common cold, are mild, but the rest are serious diseases that can disfigure, cripple, or kill their victims.

## AIDS

Acquired immune deficiency syndrome (AIDS) is caused by a retrovirus (HIV) that attacks and cripples the human immune system.
Since the disease was first described in 1981, more than 60 million
people have been infected worldwide and more than half of them
have died. The virus was isolated in 1983 by Luc Montagnier and
Françoise Barré-Sinoussi of the Institut Pasteur in Paris. In 2008,
they received the Nobel Prize in physiology or medicine for this
achievement.

HIV, a member of the *Lentivirus* genus, was transmitted to
humans from monkeys or chimpanzees in rural Africa, presum
ably through direct contact with infected primate blood. Analysis
of the HIV genome and the known rate at which it mutates suggest that the original human infections occurred in the 1930s. This
was around the time when the African population was increasing
dramatically. As a consequence, hunters began to kill and consume
greater numbers of primates, thus increasing the likelihood that a
cross-species infection would occur.

This virus cripples the human immune system by infecting
T cells, a type of lymphocyte that plays a central role in the adaptive
immune response. The human immune system is composed of a diverse group of white blood cells that are divided into three major categories, all of which are produced in the bone marrow or the thymus
gland: granulocytes, monocytes, and lymphocytes. Granulocytes
have a distinctive, lobular nucleus and are phagocytic (can eat cells
and viruses). Monocytes are large phagocytic cells with an irregu-
larly shaped nucleus. The largest monocytes, the macrophages, can
engulf whole bacteria as well as damaged and senescent body cells.
Lymphocytes have a smooth morphology and a large round nucleus.
T lymphocytes (also called T cells) and natural killer (NK) cells deal
primarily with coordinating the immune response and with killing
already infected body cells. B lymphocytes (B cells) are nonphago-
cytic; they deal with an invading microbe by releasing antibodies.

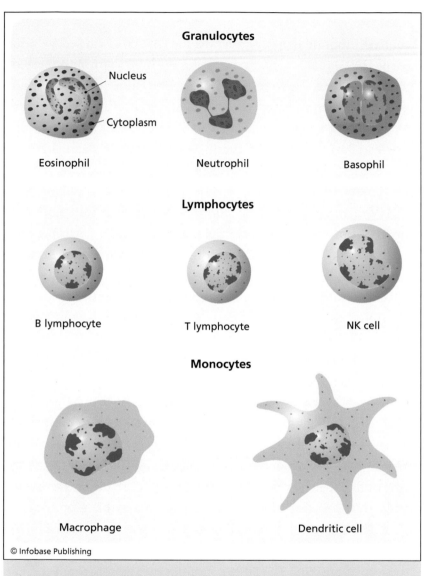

White blood cells. These cells are divided into three major categories: granulocytes, lymphocytes, and monocytes. Granulocytes have a distinctive lobular nucleus, granulated cytoplasm, and all are phago-cytic (eat cells and viruses). Lymphocytes have a smooth morphology with a large round or kidney-shaped nucleus. B lymphocytes are non-phagocytic but produce antibodies. T lymphocytes and natural killer (NK) cells can force infected cells to commit suicide. Monocytes are large phagocytes that help coordinate the immune response.

Phagocytosis of an invading microbe by granulocytes and monocytes represents a first-line defense, called the innate response. All animals are capable of mounting this kind of defense. Activation of the lymphocytes leads to a more powerful, second line of defense, called the adaptive response, which is found only in higher vertebrates. The adaptive response is initiated by monocytes, specifically, dendritic and Langerhans cells. These cells, after engulfing a virus or bacteria, literally tear the microbe apart and then embed the pieces, now called antigens, in their membrane. The bacterial antigens, bound to a group of cell-surface glycoproteins called the major histocompatibility complex (MHC), are presented to lymphocytes, which become activated when their receptors bind to the microbial antigens. Activated B lymphocytes secrete antibodies specifically designed for that particular microbe. Activated T lymphocytes and NK cells attack the microbe directly but are primarily concerned with locating and killing infected body cells.

HIV infects a special class of T cells that express two cell-surface glycoproteins: CD4, a member of the major histocompatibility complex mentioned above, and CXCR4, a chemokine receptor. Chemokines are paracrines (local-acting signaling molecules) that are released by the cells of the immune system to coordinate a response to an infectious agent. (Chemokines are discussed in greater detail in chapter 9.) HIV binds first to CD4 and then to CXCR4, which activates fusion with the cell membrane. The infection cycle of HIV was described and illustrated in chapter 2. The main function of the CD4 T cells is to help CD8 T cells that are responsible for killing virus-infected cells (these are the natural killer, or NK, cells mentioned previously). Many of the infected CD4 T cells are killed, but those that survive become resting memory cells. Recall that soon after HIV infects a cell, the RNA genome is converted to DNA, which translocates to the nucleus and is inserted into one or more of the host's chromosomes. There is little or no virus gene expression in the memory cells and consequently these cells become a stable reservoir for the virus. Monocytes and macrophages also

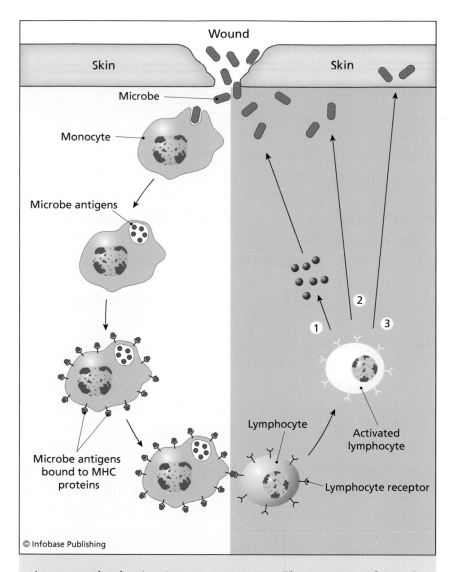

Innate and adaptive immune response. Phagocytosis of invading microbes is called the innate response (yellow zone). In higher vertebrates, microbe antigens, bound to special monocyte surface proteins called the major histocompatibility complex (MHC), are then presented to lymphocytes. Contact between the lymphocyte receptor and the antigen activates the lymphocyte and the adaptive response (blue zone), consisting of a three-pronged attack: 1) release of antibodies, which kill the microbes; 2) direct attack on the microbes; 3) destruction of infected cells.

have CD4 in their cell membranes and so are infected by HIV as well. The co-receptor in these cells is CCR5, another chemokine receptor. These immune cells do not suffer any damage due to the infection but serve as an additional reservoir that transports the virus throughout the body.

Initial symptoms of HIV infection include fatigue, rash, nausea, night sweats, and headache. Since HIV targets and destroys CD4 T cells, the long-term symptoms are associated with a severely depressed immune system. Opportunistic infections and cancers, particularly Kaposi's sarcoma, are common among those infected with HIV. Serious symptoms develop five to 10 years after the initial infection and include fatigue, malaise, fever, weight loss, shortness of breath, encephalitis (acute inflammation of the brain), and dementia. The opportunistic infections include a range of pathogens, including fungi, bacteria (e.g., *tuberculosis, Salmonella*), and other viruses (e.g., herpes, adenovirus, and hepatitis B and C viruses). Without treatment, death occurs about two years after these symptoms appear.

Despite nearly 30 years of intensive research, scientists have been unable to produce a vaccine for this disease. The main problem is the reservoir of infected memory cells that protects the virus from any therapy that is applied. For the time being, there is simply no way to remove the HIV genome from the host genome, and for this reason many scientists believe that it will never be possible to develop an effective vaccine to cure this disease.

But there is still hope. Traditional vaccines may not work, but there has been some success with drugs that block the activity of the viral reverse transcriptase, the enzyme that converts the RNA genomes to DNA, and a viral protease that is necessary for maturation of the viral particles during the infection cycle. Future strategies may be based on gene therapy. There are at least two ways to approach this form of therapy, both of which would use HIV as the chosen vector, because it is already designed to enter the target cells. The first approach would involve replacing a portion of the vector genome with a sequence designed to scramble host copies of the HIV genome (the so-called provirus). Infecting patients with this vector would lead to the reverse transcription of the engineered genome followed by its incorporation into host chromosomes, only in this case the DNA would insert within the provirus, thus scrambling or blocking the formation of any transcriptional products. The second approach would replace a portion of the vector genome with microRNA, designed to interfere with HIV RNA transcripts.

MicroRNA (miRNA), also known as small interfering RNAs (siRNAs), are a whole new class of recently discovered RNA molecules. These molecules are about 22 nucleotides long, are noncoding (i.e., do not form mRNA), and can silence a gene by binding to, thus inactivating or destroying, the gene's mRNA. The general approach is known as RNA interference (RNAi), which is currently being tested as a way to treat certain forms of cancer. This procedure shows great promise and may eventually form the core strategy of an effective AIDS therapy.

## CANCER

Human cancers are usually caused by genetic defects, which are either inherited or induced by environmental carcinogens, such as cigarette smoke, toxic industrial compounds, or ultraviolet light. But in 10 to 15 percent of the cases, cancers are known to be caused by viruses.

Cancer-causing viruses are usually DNA viruses, such as the papillomavirus (PMV), which causes cervical cancer, and the hepatitis B virus (HBV), which causes liver cancer. These two forms of cancer represent more than 90 percent of all known virally induced cancers. Characterization of PMV proteins has made it possible for scientists to produce a vaccine, which destroys the virus, thus curing the cancer. The anti–PMV vaccine is called Gardasil and was approved for medical use in 2006. Similarly, recent studies characterizing the envelope glycoproteins of HBV have led to an effective vaccine to prevent initial infection with this virus. In both cases, it is a long-standing chronic infection with these viruses that leads to cancer development. Thus, it is hoped that by blocking the initial infection it may be possible to eradicate both of these cancers.

Cancers are also caused by retroviruses; the most commonly studied is a sarcoma that occurs in chickens. In humans, retroviruses cause a rare form of T cell leukemia and Kaposi's sarcoma, which is caused indirectly by HIV owing to its ability to cripple the immune system. T cell leukemias have also been induced accidentally by gene therapy vectors. Such a case occurred during a French gene therapy trial designed to treat an inherited immune deficiency. In January 2003, two of the patients, P4 and P5, were diagnosed with T cell leukemia and, by the end of that year, researchers had determined that the cancerous cells could be traced to the genetically modified stem cells used in the treatment. It was also shown that the retrovirus used in the trial caused the cancers by inserting within or near an oncogene (a cancer-causing gene) known as *Lmo2*. Patients P4 and P5 were treated with chemotherapy and while P5 is now doing well, P4 did not respond to the therapy. Despite a second round of chemotherapy, treatment with a monoclonal antibody, and a bone marrow transplant, the child died in 2008. Shortly before the death of P4, two additional patients (P7 and P10) developed leukemia. Both patients were treated for acute lymphoblastic leukemia and after two cycles both patients, now under supportive care, are in complete remission and doing well.

Insertion of the viral genome, or portions of it, into the host genome is the most common mechanism by which viruses cause cancer. Insertion of the viral DNA damages the host DNA, which contributes to the transformation of the cell. This process is called insertional mutagenesis, but since most of the insertions occur in the intervening sequences (the safe zones) only on very rare occasions is an important host gene damaged. One such gene is called *P53*, which codes for a protein (P53) that helps regulate the cell cycle.

P53 (a lab-book name that refers to its weight) blocks progression through the S-phase of the cell cycle when DNA damage is detected (the cell cycle is discussed in chapter 10). It does this indirectly by activating the synthesis of another protein called P21, which binds to the DNA to block replication. In addition, *P53* mediates external requests (primarily from T cells) for the cell to commit suicide. Consequently, cells lacking a functional *P53* gene can divide without restraint, and they are no longer under the control of the immune system or inhibiting signals from neighboring cells, meaning they are immune to apoptosis (orchestrated cell death). Because of its protective function, *P53* is known as a tumor suppressor gene.

A second mechanism, introduced in chapter 5, is virus-mediated gene transfer. This can happen when a retrovirus transfers an oncogene to its host. Oncogenes were once normal cellular genes, called proto-oncogenes, that are sometimes picked up by retroviruses during an infection cycle. Under the influence of the viral genome, a proto-oncogene may be converted to an oncogene, and when it is transferred to a human or chicken host it becomes a potent carcinogen. Fortunately, this form of cancer induction is rare, for there are no vaccines to deal with the retroviruses that are responsible.

## CHICKEN POX

Despite its name, this disease is caused by a herpes virus called varicella and not a poxvirus. This is a relatively mild, highly contagious disease that usually affects children. The earliest clinical symptoms are malaise (the victim does not feel well) and fever, soon followed by a rash that appears first on the trunk and then on the face, limbs, and occasionally in the mouth. The rash usually lasts for five days, and most children will develop several hundred skin lesions that look like blisters. The lesions pass through four stages, beginning with macules, followed by papules, vesicles, and crusts (or scabs).

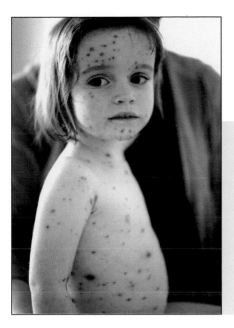

A young girl with chicken pox. This is a common infectious disease of childhood caused by the varicella-zoster virus. While the disease is not serious in children, in adults it can be dangerous. Calamine lotion can be used to relieve itching until the disease clears up. *(Ian Boddy/Photo Researchers, Inc.)*

Complications are rare in normal children, and the mortality rate, usually caused by encephalitis, is extremely low. Varicella pneumonia, which is sometimes fatal, is a common complication in adults, neonates, and immunocompromised patients who are infected with HIV or have had an organ transplant. Children with leukemia sometimes suffer extreme complications from a chicken pox infection. In some cases, varicella can form a latent provirus in the host tissues and for reasons that are unclear may be reactivated when the individual is an adult. This disease is called zoster, and it is for this reason that the virus is often called varicella-zoster. Zoster usually occurs in immunocompromised individuals or elderly adults, rarely affecting healthy young adults. It usually starts with severe pain in the skin overlying sensory nerves and ganglia. Vesicles appear within a few days, often covering the trunk, head, and neck. Zoster is often followed by protracted pain that may last for months.

A live attenuated varicella vaccine was approved in 1995 for general use in the United States. A similar vaccine has been used in

Japan for more than 30 years, but some countries, such as Canada and Britain, believe that the best way to avoid an adult outbreak of zoster is to acquire a natural resistance to the virus by being infected with varicella as children. Nevertheless, a shingles vaccine, which is a more potent version of the varicella vaccine, is available.

## COMMON COLD

The common cold is caused by the rhinoviruses *(Picornaviridae)* and to a lesser extent by the coronaviruses and adenoviruses. The incubation period is brief, and the acute illness usually lasts for seven days although a cough may persist for two to three weeks. The average adult is infected two to three times each year. The usual symptoms include sneezing, nasal obstruction, runny nose, sore throat, and headache. There is usually no fever. Secondary bacterial infections sometimes occur and may produce sinusitis, bronchitis, or pneumonitis. There are no clinical tests to distinguish between colds that are caused by the different viruses mentioned above.

There are more than 100 species of rhinoviruses, which are divided into major and minor groups depending on whether they infect a cell through the intercellular adhesion molecule (the major group) or low-density lipoprotein receptor (the minor group). This variation is largely responsible for the fact that there is currently no vaccine available for this disease.

## HEMORRHAGIC FEVER

This is a devastating disease caused by the filoviruses Ebola and Marburg. The disease first appeared in Africa in northern Zaire (now the Democratic Republic of the Congo) in 1976, where it infected 318 people with a mortality rate of 88 percent. Most of the victims lived near the Ebola River, from which the virus gets its name. Ebola reappeared in 1995 when it struck 316 people in Kikwit, in southern Zaire, with a mortality rate of 77 percent. The actual number of those infected is believed to have been much higher as there were reports of many people being infected and dying in

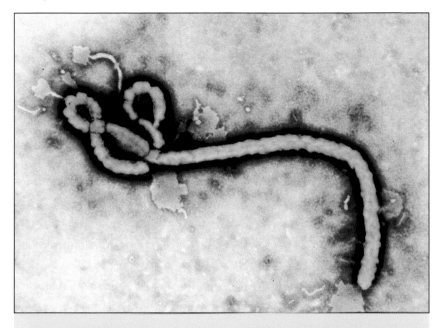

Ebola virus. The virus was named after a river in Zaire in Africa where it was first discovered in 1976. It is an RNA virus of the family *Filo-viridae* known to cause the often-fatal disease Ebola hemorrhagic fever. *(U.S. Centers for Disease Control/Frederick Murphy)*

the bush. The virus infected young and old alike, but the average victims were young adults. The Marburg virus is similar to Ebola, but it occurred for the first time in Marburg, Germany, in 1967, where it infected 31 scientists who were using African green monkeys as research subjects. Seven of the victims died. It is now known that African monkeys, chimpanzees, and gorillas are carriers of the disease, and that the reservoir hosts are rodents and bats.

Ebola can spread either through the air or by exposure to contaminated blood. It is for this reason that victims' families, who customarily accompany African patients to the hospital in order to provide food and care, become infected. The incubation period of this disease is six to 10 days, after which the patient experiences a rapid onset of fever, headache, muscle pain, weakness, conjunctivitis, and abdominal pain. The eyes develop a sunken look, and the patient becomes very lethargic. After a few days, the patient becomes nauseous and begins vomiting blood, coupled with bloody diarrhea and hemorrhaging from the mouth and nasal passages. A rash then appears, and the patient usually dies within six to nine days after the first symptoms. For the survivors, recovery can take more than a month and is associated with weight loss and profound exhaustion. The most recent outbreak of Ebola occurred in Uganda and Congo in 2007. Currently, there is no vaccine available to prevent or treat this disease. Treatment is aimed at maintaining renal function, electrolyte balance, and providing blood transfusions.

## HERPES SIMPLEX

In addition to chicken pox, discussed above, the herpes viruses are associated with infections of neurons and produce lesions (fever blisters) in and around the mouth and the genitals. The herpes simplex virus (HSV) is extremely widespread in the human population and is responsible for gingivostomatitis (cold sores in and around the mouth), encephalitis, genital disease, and infections of newborns.

There are two forms of HSV called HSV-1 and HSV-2, both of which produce skin lesions similar to those caused by varicella. HSV-1 infections are usually limited to the mouth and throat, while HSV-2 is a sexually transmitted disease and the infection remains in the lower half of the body. In either case, the virus replicates at the point of infection and then invades local nerve endings. HSV-1 infections result in latent infections in the trigeminal ganglia (a nerve group located on both sides of the head just above the ears), whereas HSV-2 infects sacral ganglia (a nerve group located near the base of the spine).

Vaccines are being developed that target the virus's envelope glycoproteins. Several antiviral drugs are available that have proved to be effective. These drugs, such as acyclovir and vidarabine, block viral DNA synthesis and thus limit the expression of clinical symptoms. The latent infections cannot be treated in this way and thus remain for the lifetime of the host. Fortunately, once the initial infection has been treated, only a small portion of the population experiences a recurrence of the disease.

## INFLUENZA

This disease, caused by the orthomyxoviruses, is very common and usually relatively mild, but in certain cases it has been known to be one of the deadliest of all viral infections. Indeed, influenza accounts for more than half of all acute diseases that occur in the United States each year. Over the past 100 years, influenza has been responsible for millions of deaths worldwide.

There are three types of influenza viruses, known simply as type A, B, and C. Type A is usually found in birds but can also infect humans, pigs, and horses. Types B and C are found primarily in humans, where they cause a mild flu that rarely leads to an epidemic or pandemic. Type A is, by far, the most deadly and is responsible for all the flu pandemics. Influenza viruses have a relatively simple structure consisting of an RNA genome that is surrounded by a membrane containing two proteins, hemagglutinin (H) and neuraminidase (N). The virus uses the hemagglutinin to enter the host cell. Naturally occurring mutations have produced 16 different subtypes of H and nine different subtypes of N. Thus, an H1N2 virus contains hemagglutinin subtype 1 and neuraminidase subtype 2 in the surface membrane. This mix-and-match approach is responsible for the annual variation in the circulating influenza strains. In addition, the exact nature of the viral subtype can have a dramatic impact on the virulence of the virus. The variation in these envelope glycoproteins also poses a serious challenge to controlling these viruses.

Influenza, which mainly attacks the upper respiratory tract, spreads from person to person by airborne droplets or by contact with a contaminated surface. The incubation period is usually two to four days depending on the initial exposure and the health of the individual. Cell destruction and the production of virions peak after 24 hours, remain elevated for a couple of days, and then decline over the subsequent five days. Initial symptoms usually include chills, dry cough, and a headache; these are followed by a high fever that can last for several days, muscle aches, anorexia, and malaise. The cough and weakness may persist for several weeks. These symptoms may be caused by any of the type A or B strains. By contrast, type C usually causes a much milder illness, very similar to a common cold.

Influenza can lead to pneumonia, but this usually occurs in the elderly and those of poor health. Pregnancy has been known to be a risk factor for lethal pneumonia in some epidemics. The secondary infection may be viral or bacterial. Influenza weakens the individual to such an extent that it increases susceptibility of the patients to bacterial superinfection. Combined viral-bacterial pneumonia is a common complication. Coinfection with the bacterium *Staphylococus aureus* can increase the fatality rate by more than 40 percent. These bacteria appear to enhance the lung tissue infectivity of the flu virus.

Tailored influenza vaccines are produced every year in an attempt to control this disease. Because it takes several months to produce the vaccines, scientists must make an educated guess as to which strain will be circulating in the upcoming flu season. Usually the guess is reasonably accurate and the vaccine successfully controls the epidemic. In other cases, when an unexpected strain begins to circulate, the vaccine proves to be ineffective. Nevertheless, annual vaccinations are highly recommended for everyone. The only people who should not get them are those who are allergic to egg protein. Since vaccine strains are grown in eggs, residual egg proteins are usually present in the vaccine.

## MEASLES

Measles, caused by the paramyxoviruses, is the most common contagious disease among children. The WHO has estimated that

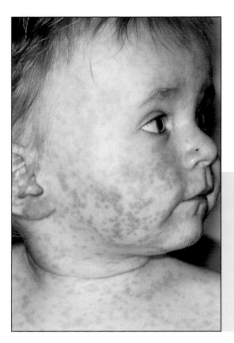

Measles rash on a child's face. Measles is a highly infectious viral disease. It mainly affects children but can occur at any age, and the adult form is more severe. An attack usually gives lifelong immunity. *(P. Marazzi/ Photo Researchers, Inc.)*

infections caused by the paramyxoviruses are responsible for the deaths of more than 4 million children worldwide each year.

The infection begins in the respiratory tract and after a two-week incubation period spreads to the skin and lymphoid tissue. Measles (from middle English "maseles," which means pustule) is associated with a red to brown rash, high fever, and conjunctivitis. The virus can replicate in certain lymphocytes, which spreads the virus throughout the body and in some cases the infection will spread to the central nervous system (CNS). The conjunctivitis is commonly associated with an aversion to light. Pneumonia is the most common complication and is usually caused by a secondary bacterial infection. Complications involving the CNS are the most serious, appearing as acute encephalitis in about 1:1,000 cases, with a mortality rate of 15 percent. Overall, the mortality rate in healthy individuals is about 0.3 percent but can be as high as 30 percent among the malnourished.

A highly effective attenuated live virus vaccine has been available since 1963. The vaccine has reduced the annual rate of measles in the United States from 500,000 cases in the 1950s to less than 40 in 2009. Studies have shown that vaccine-induced antibodies persist for up to 33 years and thus the vaccine is assumed to give a lifelong immunity.

## POLIO

Poliomyelitis has crippled humans for thousands of years. This disease is caused by the picornaviruses (literally, tiny RNA viruses), some of which are known as enteroviruses because they infect by way of the digestive tract. Enteroviruses attack many animals including cattle, pigs, mice, and monkeys. Infection in cattle is called foot-and-mouth disease. Polioviruses are restricted to primates. The absence of a primate-specific cell-surface receptor makes other animals resistant to this disease.

Polioviruses enter the body through the mouth, and the initial infection involves cells in the throat and intestines. The incubation period may last for two weeks, during which time the virus is absent from the throat but appears in abundance in intestinal cells, even though high antibody levels are present in the blood. The virus can also be found in the tonsils and the lymph nodes of the neck. In serious cases, the virus infects the CNS by way of the circulatory system. The virus can also spread along axons of the peripheral nerves to the CNS to involve neurons in the spinal cord and the brain. Replication of the virus destroys neurons, primarily along the spinal cord and peripheral nerve fibers. Death of these cells produces the paralysis that is characteristic of this disease.

The course of this disease ranges from subclinical (no signs of illness), mild febrile illness, to severe permanent paralysis. The mild form is the most common and is characterized by a fever, headache, nausea, vomiting, sore throat, and constipation. The patient usually recovers in a few days. A slightly more serious form of the disease, called aseptic meningitis, occurs in some cases. The symptoms include all of those described for the mild form in addition to pain

Young polio victims. Eleanor Roosevelt greeted these children in Washington, D.C., on January 22, 1940. *(Science Source/Photo Researchers, Inc.)*

in the back and neck. Recovery from this form of the disease also takes but a few days. Paralytic polio occurs in about 1 percent of the cases. In some cases, neural control of the respiratory system is damaged, and the patient requires the "iron lung" to force the exchange of gases into and out of the lungs until the nerves recover. Paralysis is caused by the destruction of motor neurons branching off the spinal cord and in some cases the damage can reach as high as the brain stem. The amount of damage varies greatly, with some patients showing a full recovery after six months while others are permanently paralyzed. Polio does not affect muscle tissue directly but the withered arms and legs that are characteristic of this disease are due to degeneration of the muscle as a result of having lost the innervation. In addition, a withered limb is especially pronounced when the individual contracts the disease as a child, since the growth of the limb is permanently retarded.

An effective vaccine has been available since 1955, and the WHO has launched a major campaign to eradicate this disease. Prior to vaccination, there were about 21,000 cases of paralytic polio every year in the United States. Currently, the Americas are free of the disease, but much work is needed to eliminate it from undeveloped countries.

## RABIES

Rabies (from the Latin "rabere," meaning to rave) is caused by the rhabdoviruses. This disease is an acute infection of the CNS that is nearly always fatal. It is primarily a disease of lower animals, which is transmitted to humans from the bite of an infected animal, such as a dog, bat, or raccoon. Indeed, the vampire bat is known to transmit the disease for months without ever showing signs of the disease. The virus has a wide host range, and nearly all mammals seem to be susceptible. The disease was common in Europe during the 19th century but is relatively rare today.

The virus multiplies in the tissue at the source of the infection. The incubation period in humans may be one or two months, after which it spreads up the nerves to the CNS. Progressive encephalitis develops as the virus multiplies in the brain cells. It eventually spreads through peripheral nerves to the salivary glands and other tissues. The migration to the salivary glands is an effective adaptive strategy for the virus since the animal's saliva becomes infectious. The host can then spread the virus around by licking or biting another animal. During the acute neurologic phase, which can last up to one week, patients show many signs of CNS damage, characterized by nervousness, hallucinations, and bizarre behavior including a morbid fear of water. The act of swallowing initiates a painful spasm of the throat muscles. This phase is followed by convulsive seizures, coma, and death. The major cause of death is respiratory paralysis. The full course of the disease can last for a month.

The first rabies vaccine was produced by Louis Pasteur in 1885, and there are many effective forms available today. Although very few cases of rabies occur in the United States each year, more than 20,000 persons receive some form of preventive vaccination. The decision to administer this treatment depends primarily on the severity of the bite, the identity of the animal, and whether it is available for testing. Worldwide, more than 50,000 people die of rabies every year. Most of these cases occur in Asia and Africa.

## SARS

Severe acute respiratory syndrome (SARS) is a disease that first appeared in 2003 and is now known to have been caused by a coronavirus. Normally, coronaviruses infect the upper respiratory tract and, like the rhinoviruses, produce the common cold.

But the SARS strain produced a serious respiratory ailment, including pneumonia, which led to respiratory failure. The incubation period was about six days and was followed by fever, chills, headache, cough, dizziness, sore throat, and shortness of breath. In many cases, the patient progressed rapidly to acute respiratory distress and had to be put on a ventilator. The disease had a mortality rate of about 10 percent.

There are no vaccines available for this disease. SARS was contained and presumably eradicated by isolation of patients, quarantine of those who had been exposed, travel restrictions, and the use of respirators by hospital personnel. The SARS pandemic will be discussed in the next chapter.

## SMALLPOX

The smallpox virus (variola) is an ancient human pathogen that killed millions of people worldwide until it was eradicated in the 1970s (estimates range as high as 500 million deaths). There are also several species in the *Poxviridae* family that infect monkeys, cows, camels, and many other animals. Although eradicated, there is a persistent fear that the few remaining laboratory stocks of this virus could be accidentally or intentionally released into the environment, leading to a new wave of epidemics.

Variola enters the body through the upper respiratory tract (the mouth and nose). The incubation period is about two weeks, during which time the virus multiplies in the lymphatic tissue adjacent to the point of entry and then spreads, via the circulatory system, to the spleen, lymph nodes, liver, and lungs. The onset of clinical symptoms is rapid and is associated with fever, abdominal pain, vomiting, headache, and prostration (too sick to stand). The viruses then infect the skin, producing the sores that are characteristic of this disease. The skin lesions pass through the same stages observed with chicken pox, but in this case the staging is synchronized, that is they all reach the final crusting stage at about the same time. Initially, the lesion is but a spot on the skin (macule), which then progresses to a raised lesion (papule) that fills with fluid (vesicle). After two weeks of infection, the vesicles are transformed into pustules, which eventually dry out to form the crusts (scabs). The lesions are most abundant on the face and less so on the trunk. White scars of varying severity are revealed as the crusts fall off. In some cases the scars fade over time, but most of the survivors are

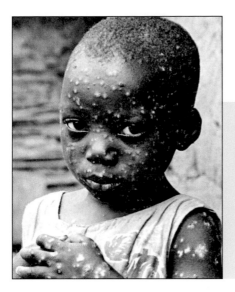

A Congolese child with smallpox. Smallpox has now been eradicated from the world as an infectious disease, and only laboratory workers or military members are vaccinated for it. The photo was taken in 1965. *(Science Source/Photo Researchers, Inc.)*

disfigured with permanent scars. Mortality is about 30 percent and is due to a massive inflammatory immune response causing shock and multiorgan failure, which usually occurs in the second week of illness.

The British physician, Edward Jenner, produced an effective smallpox vaccine in 1796 (see chapter 4). And yet it took 184 years to eradicate this terrible disease. The reasons for this extreme and most unfortunate delay are complex, involving social acceptance of vaccination and the political will to initiate and maintain a vaccination program. These issues will be discussed in the next chapter.

## WEST NILE FEVER

This disease is caused by a flavivirus. It occurs primarily in Africa, the Middle East, Southeast Asia, and more recently in the United States. Its first appearance in North America occurred in 1999 when it infected and killed several exotic birds in New York City. Within three years it had spread throughout the country where infected humans, birds, and horses.

The virus currently infects from 3,000 to 4,000 people each year and is the leading cause of encephalitis in the United States. More than 80 percent of the infections are subclinical, and less than 1 percent lead to meningitis or encephalitis. This virus can pass from person to person, and there are several cases where transmission occurred through blood transfusions and organ transplants. The high rate of subclinical infections suggests that this disease is serious primarily for those who are immunocompromised. This group includes those individuals who have had an organ transplant, are receiving cancer therapy, or are suffering from an HIV infection.

## YELLOW FEVER

Yellow fever is a mosquito-borne disease that is caused by members of the *Flaviviridae* family. The disease is believed to have been introduced to the Americas in the 17th century with the importation of African slaves.

The incubation period is only a few days, after which the victim develops a high fever, headache, chills, and dizziness. These early symptoms are quickly followed by nausea, vomiting, yellowing of the eyes, and bradycardia (abnormally slow heartbeat). In most cases, the patient recovers after a few days, but in about 15 percent of cases the infection seriously damages the liver, kidneys, spleen, and heart. Damage to the myocardium (heart cells) leaves the patient in shock. When the disease progresses to this stage, the mortality rate can be more than 20 percent. The cause of death is due to internal bleeding and multiorgan failure (heart, liver, and kidneys).

In the 1890s, scientists working for the U.S. Yellow Fever Commission, under the direction of Walter Reed, an American physician, showed that the disease was transmitted by the mosquito *Aedes aegypti* and that the pathogen was a virus. In 1937, Max Theiler, a British scientist working at the Rockefeller Foundation in New York, successfully attenuated the yellow fever virus, thus producing a strain (17D) that was used to develop an effective vaccine. He received the Nobel Prize in physiology or medicine in 1951.

## SUMMARY

Viruses are responsible for at least 65 human diseases. While some of these diseases, such as the common cold and certain forms of influenza, are relatively mild, the rest are serious diseases that can cripple or kill the human victim. In the past 100 years, viral diseases have killed millions of people. Smallpox alone has killed more than 300 million people, and by some estimates measles and influenza have killed another 200 million. Viruses, unlike bacteria or fungi, attack a cell at its most fundamental level, commandeering basic cellular and genetic functions for its own purposes. Viruses can infect the lungs, muscles, nervous system, circulatory system, skin, and attack and cripple the immune system.

# Viral Pandemics

Viral diseases are rarely, if ever, contained within a single geographical location. This is especially true today with airlines providing quick and easy access to just about every point on the globe. Because air travel is so quick, people carrying a deadly viral infection may embark on a journey and arrive at the destination before they realize they are sick. By that time, they have already passed the virus on to their fellow travelers and anyone else they encounter at their destination. This chapter discusses six pandemics, the first four of which have killed many millions of people worldwide. The rest, although relatively mild diseases, serves to illustrate some of the problems associated with the control of diseases. Organizations that coordinate the battle against infectious diseases, such as the World Health Organization (WHO) and the U.S. Centers for Disease Control (CDC), have to strike a fine balance between the need to sound the alarm and the danger of scaring people when the threat is mild. Moreover, they can only work effectively when reports of epidemics from around the world are reliable.

## AIDS

HIV/AIDS was described as such for the first time in Los Angeles, California, in 1981. The patients were four homosexual males who had been hospitalized for prolonged bouts of fever and multiple bacterial, fungal, and viral infections. This new disease reached epidemic proportions by 1983, and since that time it has spread around the world, infecting more than 60 million people and killing nearly half of them. Antiviral drugs have reduced the mortality rate, but worldwide nearly 58 percent of those afflicted with this disease never receive treatment of any kind. Currently, there are more than 33 million people living with HIV. In 2009, 2.6 million new cases and 1.8 million deaths were reported worldwide.

Although the first patients diagnosed with AIDS were Americans, the disease did not originate there nor is it confined to the homosexual community. DNA analysis has traced the source of HIV to African chimpanzees, *Pan troglodyate,* living in southern Cameroon. Scientists believe that people in that area became infected with HIV by hunting and eating those animals. Recent estimates suggest that the first infections occurred around 1908 and that the earliest documented cases of HIV, and the source of the pandemic, occurred in 1959 near Kinshasa in the Democratic Republic of the Congo. Soon after these cases were reported, HIV infections began to appear in Haiti, Europe, and the United States. Several factors have been cited as the reason for the spread of this disease: an increase in international travel, increased sexual promiscuity, increased use of blood transfusions, and a marked increase in intravenous drug use. As the epidemic grew, there was a suggestion that HIV was introduced to the African continent by a contaminated polio vaccine. After careful examination by several international commissions, it was shown that viruses present in the tissues of chimpanzees used to make the polio vaccine are distinct from HIV, and that the original cell cultures used to make the vaccine were not infected with HIV.

A recent United Nations report on the AIDS pandemic (UN-AIDS, 2010) shows that the disease is now prevalent in nearly every region of the world. The number of cases ranges from 57,000 in Oceania (Pacific islands, Australia, New Zealand) to just more than 22 million in sub-Saharan Africa. For nearly every region, the number of cases has increased since 2001, but there are signs that the numbers are beginning to level off.

Although HIV is transmitted primarily by sexual intercourse and intravenous drug use, contaminated blood supplies have been a major source of infection. During the early 1980s, thousands of

## HIV/AIDS CASES AROUND THE WORLD

| LOCATION | 2001 | 2009 |
|---|---|---|
| Sub-Saharan Africa | 20.3 million | 22.5 million |
| Middle East and North Africa | 180,000 | 460,000 |
| South and Southeast Asia | 3.8 million | 4.1 million |
| East Asia | 350,000 | 770,000 |
| Oceania | 29,000 | 57,000 |
| Latin America | 1.1 million | 1.4 million |
| Caribbean | 240,000 | 240,000 |
| Eastern Europe and Central Asia | 760,000 | 1.4 million |
| Western and Central Europe | 630,000 | 820,000 |
| North America | 1.2 million | 1.5 million |

Note: Table values are for adults and children living with HIV. Data is from the 2010 United Nations Report on HIV.

hemophiliacs were given blood that was contaminated with HIV. Even after 1985, when researchers at NIH developed a way to test the blood supplies for HIV contamination, untested blood was being given to unwary hemophiliacs. This happened in the United States, France, and Germany in the late 1980s, and in Japan in 1996. Several lawsuits were launched and key figures in the blood-supply industry and in the Japanese government were charged with many counts of murder (ranging from 300 in the United States to 6,000 in Germany) for selling blood that was contaminated with HIV. Despite these notorious trials and the existence of an accurate test, HIV-contaminated blood was still a problem as late as 2007. In that year, Dr. Gao Yaojie, a Chinese physician, was placed under house arrest in Zhenzhou for announcing to the world that HIV-contaminated blood was being used in rural areas of Henan, a province in the east central part of the country, and was largely responsible for the spread of the disease in that province and throughout China. Dr. Gao was arrested to prevent her from receiving an award in the United States for her work in caring for AIDS patients at a time when the government was denying the existence of an AIDS epidemic in the country. The Chinese government has since changed course and is now providing funds to fight the disease.

Dealing with the AIDS pandemic has been especially difficult in Africa where the great majority of cases are to be found. The main problem is the small number of health care providers; in some places there may be only one physician for 40,000 people. There are also the difficulties associated with trying to deliver antiviral drugs to inaccessible communities or communities that are in the middle of a war zone. Nevertheless, according to the UNAIDS report, antiviral therapy coverage in Africa rose from barely 7 percent in 2003 to 37 percent in 2009. Overall, 5.2 million people in low- and middle-income countries are receiving therapy, which amounts to a tenfold increase over the past five years. The report makes the following recommendations for fighting the AIDS pandemic in the future:

1.  Reduce sexual transmission of HIV.
2.  Prevent mothers from dying and babies from becoming infected with HIV.
3.  Ensure that people living with HIV receive treatment.
4.  Prevent people living with HIV from dying of tuberculosis.
5.  Protect drug users from becoming infected with HIV.
6.  Remove punitive laws, policies, practices, stigma, and discrimination that block effective responses to AIDS.
7.  Stop violence against women and girls.
8.  Educate young people to protect themselves from HIV.

Although these are all admirable goals, they will be extremely difficult to implement. Reducing the sexual transmission of HIV has been difficult enough in developed countries where people have ready access to condoms, but it has proven to be an almost impossible goal in less developed countries, particularly in Africa, where the men simply refuse to use protection of any kind.

The problem of tuberculosis coinfecting AIDS victims is extremely serious. HIV depresses the immune system and thus allows *Mycobacterium tuberculosis* to infect the lungs. The Bacille-Calmette-Guérin tuberculosis vaccine, developed more than 80 years ago, is not fully effective, but it is the only treatment available. Unfortunately, it is now clear that HIV-infected children are especially susceptible to side effects from the BCG vaccine. The reaction can be quite serious and is sometimes fatal.

Removing punitive laws and policies is a hotly debated topic at the present time. No one wants to be stigmatized and certainly no one should be discriminated against for having contracted a deadly disease. But the CDC, in an effort to block the transmission of HIV, has recommended that all teens and adults in the United States be tested for HIV as part of a routine medical exam. But this proposal has met with almost universal opposition from civil liberty groups and gay activists. Moreover, health insurers have refused to pay for the testing. Some hospitals do offer the tests, which are funded by the state governments or private foundations.

In some cases, U.S. courts have ruled that individuals who lead high-risk sexual lifestyles are responsible for knowing whether or not they are infected with HIV and for informing their partners about possible exposure. New York, for example, sent Nushawn Williams to prison in 1997 for 12 years for having sex with several women even though he knew he was infected with HIV. This case led to the passage of a law requiring doctors and laboratories to report to the state the names of people who test positive for HIV. Florida passed an HIV-specific law in 1997 and is currently prosecuting Darren Chiacchia, an Olympic bronze medalist, for exposing his sex partner to the virus.

Catherine Hanssens, executive director of the Center for HIV Law and Policy, insists that "these laws always were a bad idea, and the fact that they were a bad idea only becomes increasingly obvious the more we know about HIV and how it is transmitted." But James Subjack, a former district attorney in Chautauqua County in upstate New York where Williams was prosecuted, argues that such laws are necessary: "Without HIV-specific laws, how would you convince a jury, for example, that when some guy is having random sex with somebody else, that it was his intent to kill them as opposed to having sex?"

No other disease has caused this type of controversy. Certainly victims of smallpox, yellow fever, and a few other diseases were quarantined, but the people suffering from those diseases were never stigmatized as AIDS victims have been. This has happened because the earliest cases occurred in the homosexual community. Even now when it is clear that this disease can infect anyone, the problem still exists. The fear of being stigmatized is so strong that it is hampering the efforts of the CDC and other health organizations to control the transmission of HIV, and this problem will not be resolved until an effective vaccine is produced.

## POLIO

Polio is another ancient disease that has killed or crippled millions of people. Although it has been around for centuries it was not defined as a specific disease until the 17th century. It then appeared sporadically throughout the 18th and 19th centuries, when it infected people primarily in Europe, the Middle East, and North America.

The first documented polio epidemic in the United States occurred in 1894 in Rutland, Vermont. Charles Caverly, of the Vermont State Department of Public Health, described an epidemic that affected 123 people. The majority of those afflicted (68 percent) were children under the age of six, and of the total 18 died and 50 were permanently paralyzed. Polio struck again in 1905, this time in Stockholm, Sweden, where 1,200 cases were reported. Throughout the first half of the 20th century, many American cities, including New York City, reported epidemics affecting thousands of people. Franklin Delano Roosevelt, the future president of the United States and pivotal leader throughout the Great Depression and the Second World War, was stricken with the disease in 1921 at the age of 40. His wife, Eleanor Roosevelt, became a powerful force in the effort to control and eradicate this disease. The total number of cases in the United States at the time was about 40,000.

By the time Roosevelt was stricken, it was already well known that polio was caused by a virus. Researchers at the Institut Pasteur had provided proof of a viral agent in 1909 and went on to recover the virus from the throat, tonsils, lymph nodes, and intestines of polio patients who had died. Because of Pasteur's earlier successes, the public was confident that he and his coworkers would be able to produce a vaccine. But it did not happen for more than 30 years, and it was Jonas Salk, an American physician, rather than scientists at Pasteur, who produced the first effective polio vaccine.

Salk's vaccine was produced in 1952 by inactivating (killing) the virus with formaldehyde. The vaccine was tested in 1954 in a clinical trial that enrolled 650,000 children from 44 states, of whom 440,000 received the vaccine and 210,000 received a placebo. Salk's vaccine required a needle inoculation, and the clinical trial to test it is still the largest medical experiment ever conducted. The report, which followed two years later, showed that the Salk vaccine was both safe and effective, and it was quickly licensed for general use. Despite its effectiveness, there were serious problems with it initially. Although millions of doses of the vaccine were safely administered, a faulty batch of the vaccine, which had not been properly prepared and thus contained live virus, was produced. When this batch was used the unthinkable happened: 204 recipients, all of whom were children, developed polio; 153 of the victims were paralyzed and 11 died.

This terrible tragedy was the driving force behind the search for a better vaccine, which was produced in 1956 by Albert Sabin, a Polish-American physician and virologist. This vaccine had several advantages over the Salk vaccine: first, it was a live attenuated vaccine that conferred a more effective immunity; second, the preparation of the vaccine was such that the risk of it causing polio was much less than that of the Salk vaccine; and finally, because the natural route of infection is through the digestive tract, an attenuated vaccine can be taken orally so that mass immunizations are much more convenient than those using the Salk vaccine. The Sabin vaccine is administered by placing one or two drops of the viral preparation on a sugar cube, which is then consumed by the recipient. In 1960, the surgeon general announced that the Sabin vaccine would replace the Salk vaccine in the United States, and it quickly became the vaccine of choice throughout the world.

By 1996, polio was all but eliminated from North America and Europe, and more than 400 million children were immunized worldwide. But although the WHO, the United Nations Children's Fund (UNICEF), the CDC, and the Bill and Melinda Gates Foundation have continued to sponsor immunization programs, the

Polio immunization. A 1962 aerial shot of a queue of people outside an auditorium in San Antonio, Texas, being used as a polio immunization center. This site was using the oral polio vaccine developed by Albert Sabin. *(Public Health Image Library/U.S. Centers for Disease Control)*

disease has not been eradicated. The problem is due mainly to the political climate in the world today. Some countries, such as Congo, Pakistan, and Afghanistan are difficult or simply too dangerous to enter. Other countries, such as Nigeria with a large Muslim population, have questioned the need for polio vaccinations and have even claimed that the vaccine is being used by its enemies to spread HIV and to sterilize Muslim women. Bill Gates, cofounder of the Gates Foundation, encountered this type of resistance in 2010 when he visited Nigeria. But the Sultan of Sokoto, the spiritual leader of Nigeria's 70 million Muslims, offered his support in the fight against polio. With the cooperation of Nigeria and other African countries, the WHO is confident that polio will be eradicated by 2012.

## SMALLPOX

Although Edward Jenner produced an effective smallpox vaccine in 1793, the disease continued to infect and kill people all over the world for the next 184 years. This would seem to be an example of supreme irresponsibility, and to a great extent it is, but in the 18th and 19th centuries there were no organizations available that could coordinate a massive immunization program. In addition, while no one wanted to get the dreadful disease, there seemed to be a general sense of apathy toward a comprehensive vaccination program. This curious attitude may be explained by considering the design of the ocean liner *Titanic,* specifically regarding the inadequate number of lifeboats that were provided. The rule at the time seems to have been as follows: Make sure there are enough lifeboats onboard for the first-class passengers, but do not worry about anyone else. This decision condemned hundreds of second-class and third-class passengers to their deaths when the ship sank in 1912. A class-based mentality such as this could be the reason why governments in Europe and the United States were generally in favor of vaccination but could not seem to muster the resolve or support to see it through. Politicians in those days might have been willing to support an immunization program for certain members of society, but the thought of doing it for the sake of all humanity was an alien concept.

Jenner and Thomas Jefferson both expressed the hope that smallpox would someday be eliminated, but a concerted effort was not made until 1950, when the Pan-American Sanitary Organization set out to rid the Americas of this disease. Massive immunization programs were begun, and by the 1970s Argentina, Brazil, Colombia, and Ecuador were the only countries still afflicted with smallpox. In 1953, Dr. Brock Chisholm, the first director general of the WHO, proposed to expand the anti-smallpox effort into a global effort, and he asked the member states for their support. At the time, the WHO was preoccupied in its effort to eradicate malaria, a program that was in itself very complicated and very expensive. Consequently, the assembly rejected Chisholm's plan. Five years later, the USSR reported that it had successfully eradicated smallpox, and its vice minister of health Victor Zhadnov suggested to the WHO that if the Soviet Union could eliminate the disease from such a vast territory that other countries could surely do the same. Zhadnov proposed a 10-year program to fight smallpox, and this time the assembly accepted the plan but allocated only $100,000, an amount that was hopelessly inadequate and doomed the effort to failure. Despite the success in the Americas and in the USSR, many members of the assembly felt that eliminating smallpox, or any other pathogen, was unrealistic and perhaps even impossible.

Fortunately, not everyone agreed with this sentiment, and in 1967 the WHO launched another global initiative, with a budget of $2.4 million, to eradicate smallpox. At that time, there were 33 countries with endemic smallpox that infected 10 to 15 million people every year. By 1976, the disease was eliminated from Africa, Brazil, Indonesia, and the entire Asian continent. The last case of smallpox occurred in Somalia in 1977. The WHO announced the worldwide eradication of this disease in 1980.

## SPANISH FLU OF 1918

During winter 1918, just as World War I was drawing to a close, Europe was hit with the worst outbreak of influenza in recorded history. This deadly disease, known as the Spanish flu, quickly became a pandemic as it spread to other countries around the world. In just one year, this flu killed more than 50 million people worldwide, far exceeding the number of casualties from the war or from other notorious epidemics, such as the bubonic plague that swept through Europe in 1347. Indeed, the Spanish flu pandemic is now believed to have caused nearly half of the casualties of World War I.

This pandemic is known as the Spanish flu because the first reports of the outbreak in Europe happened to be from Spain, but the virus is now known to have originated in Asia, possibly China, where it was endemic in poultry and pigs. Avian (bird) flu viruses, such as the H5N1 strain, usually do not infect people directly. Pigs, on the other hand, are susceptible to avian and human influenza viruses. Thus, several strains of viruses can infect pigs simultaneously, and if these viruses recombine genetically they can produce a novel and possibly virulent new strain that is capable of infecting humans. For example, recombination of an avian type A H5N1 virus with a swine H1N2 virus is believed to have produced the virulent H1N1 strain that caused the Spanish flu.

Subsequent influenza pandemics, occurring between 1957 and 2004, were also caused by avian/swine flu viruses that originated in Asia (China, Hong Kong, Vietnam, Cambodia, and Thailand). Rural Southeast Asia is the most densely populated area in the world. In addition, it is a place where hundreds of millions of people live and work in very close contact with pigs and ducks. Under these conditions, viruses pass easily from host to host thus increasing their range and their mutation rate. Thus, it is no surprise that virulent viral strains usually originate in that part of the world. The regular appearance of new influenza subtypes poses a serious health threat throughout the world, for it is only a matter of time before one appears that is as virulent as the Spanish flu. The CDC has estimated that such an outbreak would kill up to 200,000 Americans, leaving 700,000 hospitalized and 50 million people sick, at a total cost of $150 billion. These estimates are based on the availability and rapid distribution of antiviral medications and vaccines.

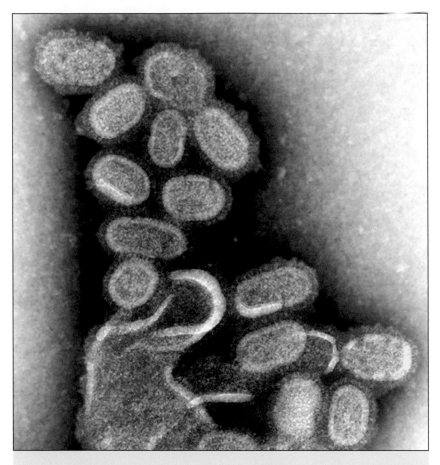

The 1918 influenza virus. This negative stained transmission electron micrograph (TEM) shows recreated 1918 influenza virions that were collected from the supernatant of 1918-infected canine kidney cell culture. *(Public Health Image Library/U.S. Centers for Disease Control)*

It is not clear why the Spanish flu virus was so deadly. Scientists believe that a subtle, though as yet uncharacterized, change in the neuraminidase was responsible for its virulence. The virulence of other influenza viruses is also believed to be a function of their subtype. An H2N2 virus initiated the flu pandemic of 1957, which killed 70,000 people in the United States alone. The Hong Kong flu of 1968 was caused by an H3N2 virus that killed 34,000 Americans. Recent outbreaks of avian flu in Vietnam and other Asian countries are caused by an H5N1 virus that has spread to humans, possibly directly from birds. Fortunately, this particular H5N1 subtype is proving to be less lethal than previously encountered avian flu viruses.

It is to be hoped that governmental organizations and the WHO will respond quickly to any future influenza pandemics. Indeed, the death toll from the Spanish flu would have been much less if international health organizations had existed in those days. The current guidelines for managing flu pandemics are as follows:

1. Close all schools for three months.
2. Close churches, theaters, and areas of assembly.
3. Employees should work staggered hours to reduce crowding on public transport.
4. Cancel athletic events.
5. All patients should be quarantined, and visiting friends and family should wear face masks.

Despite these measures, the best way to deal with influenza pandemics is to prevent them from occurring. To this end, the CDC and the WHO have undertaken a large number of studies in order to gain a better understanding of influenza viruses, the molecular nature of their virulence, and ecological questions concerning the life cycle of such viruses in the wild and on pig and poultry farms. This information may make it possible to stop the spread of virulent flu viruses even before they infect humans.

## SARS

Severe acute respiratory syndrome (SARS) was caused by a coronavirus. Normally these viruses cause the common cold, but an unusually lethal strain appeared in China in 2002. Within a year, this disease spread to more than 33 countries on five continents where it infected more than 8,000 people, 818 of whom died. The pandemic was officially contained by the end of 2004.

Although relatively mild, this is an example of a pandemic that should never have happened. The original outbreak of this disease occurred late in 2002 in Guangdong, a province located on the southern coast of China. Initially the cases were confined to people working or eating in restaurants and certain marketplaces. DNA analysis traced the source of the virus to the masked palm civet, a catlike animal used in Chinese cuisine. Subsequent analysis showed that the horseshoe bat, and not the civet, was the reservoir of the SARS coronavirus. The bats apparently infected the civets, which in turn passed it on to a few humans who infected other members of their community. Since the source of the disease was quickly identified by Chinese physicians, the question is how did SARS become a global pandemic? After the initial outbreak, the Chinese government reported to the international community and to the WHO that the disease was fully contained and that no new cases had emerged.

Unfortunately, this report was untrue and is now known to have been a complete fabrication by members of the Chinese government. In fact, in 2003 the disease was out of control and was spreading quickly throughout China. The international community would have learned nothing about this if it were not for Dr. Jiang Yanyong, a very brave Chinese military physician who provided documented proof to Western journalists that his government was trying to cover up the full extent of the epidemic. Initially, Jiang was placed under house arrest, but eventually the Chinese government was forced to admit that their reports concerning the epidemic were false. The delay in reporting the full extent of this epidemic meant that other countries were not on guard, and as a consequence this disease spread throughout the world, killing hundreds of people before the deception was exposed. In 2007, Jiang was selected by the New York Academy of Sciences to receive the Heinz R. Pagels Human Rights of Scientists Award, but the Chinese government placed him under house arrest to prevent him from collecting it.

The SARS pandemic highlights several problems that need to be dealt with in order to improve the control of viral epidemics and pandemics. The first, and most obvious, is the need for honesty and transparency whenever a potential threat is reported to the international health community. Second, greater care must be taken with novel infectious agents. In 2004, 13 laboratory workers were infected with SARS because the containment facilities were inadequate. Third, emergency procedures need to be established in order to control and analyze a new pandemic. An effective tracking system would be part of these procedures. When SARS spread to Toronto, Canada, in 2003, all but three of the 225 cases were eventually traced to a single case in Hong Kong. Such information, when it is available quickly, can go a long way to blocking the spread of a disease.

## SWINE FLU

In 2009, an influenza pandemic occurred that was caused by a type A (H1N1) viral strain (the official designation is pandemic [H1N1] 2009). Farmworkers in the United States caught this virus from their pigs, after which it spread rapidly from person to person. By the time the WHO declared a pandemic in June 2009, the virus had spread to Mexico, Canada, and 69 other countries around the world. Currently, most countries in the world have reported infections from this new virus.

Genetic analysis of the virus showed that it originated from animal (swine) influenza viruses and is not related to the human seasonal H1N1 viruses that have been in general circulation among people since 1977. The new virus produced unusual patterns of death and illness. Most of the deaths caused by swine flu occurred among younger people, including those that were otherwise healthy. Pregnant women, children, and people of any age with certain chronic lung diseases had the highest risk of being killed by this virus. Many of the severe cases were caused by viral pneumonia, which is more difficult to treat than bacterial pneumonias.

The WHO was especially concerned about this virus because it was a new H1N1 strain that had the potential for being as virulent as the Spanish flu. The public was advised to take the necessary precautions and that a vaccine would be prepared as quickly as possible. But the vaccine took longer than expected because this particular strain of H1N1 did not grow well in the chicken eggs required for vaccine production. The lag in the availability of the vaccine began to strain the public's faith in the CDC and the WHO. As the vaccine became available it was given first to children, teenagers, and the elderly. By the time there was enough vaccine for everyone the pandemic had already begun to wane, and nearly half of the adult American public had decided they did not need the vaccination after all. Moreover, there was never full agreement among the health care community regarding the danger of the virus. On one hand, the WHO had given the impression that A (H1N1) 2009 was the second coming of the Spanish flu, while Dr. Anthony Fauci, a virologist at the National Institutes of Health, was suggesting that the disease was just another seasonal flu. And indeed, subsequent analysis has shown that he was right. By 2010, swine flu had killed 16,000 people worldwide, but a typical flu season is associated with 35,000 deaths, more than twice the mortality rate of swine flu.

Consequently, the WHO was sharply criticized by the Council of Europe's health committee, which suggested that the organization had exaggerated the dangers of the swine flu and had caved in to pressure from pharmaceutical companies that profited handsomely from the 200 million doses of vaccine that were eventually produced. The WHO has defended itself by pointing out that the criticism would have been much worse if they had done nothing and the virus had turned out to be as dangerous as they believed it to be. Keiji Fukuda, a WHO virologist, insisted that the organization did what it had to do in order to protect the public from a potential threat.

The swine flu of 2009 is a good example of how difficult the management of pandemics can be. The WHO, the CDC, and other health organizations were forced to make costly decisions based on imperfect projections. If the swine flu turned out to be only half as virulent as the Spanish flu, the pandemic could have killed millions of people. Perhaps the $400 million price tag for a vaccine that few people needed is the price for an effective anti-pandemic strategy. It did not work out well this time around, but when it comes to viral diseases no one knows what the future will bring.

## SUMMARY

Pandemics occur when an epidemic gets out of control, either because the affected country lacks the resources to control the disease or because it does a poor job of alerting the international health community. Even with the greatest of care, however, rapid air travel makes containment extremely difficult. In some cases, a single infected traveler is enough to start a pandemic. Screening travelers at airports in order to avert the spread of a disease has been attempted in some cases, but controlling pandemics is a multistep process that requires honesty and diligence from countries around the world. Organizations, such as the WHO and the CDC, have established procedures for dealing with pandemics, including the coordination of vaccine production and distribution. The decisions being made by these organizations have been honed to a fine art as they try to keep the international communities alert to epidemics and potential pandemics, without inducing excessive fear and panic.

# Fighting Viral Infections

Viruses have special proteins, usually two or more, embedded in their outer coat or capsid that are used to infect a cell. These proteins bind to specific cell-surface receptors, and if the match is just right the cell responds by letting the virus in, usually by a process known as receptor-medicated endocytosis. But these viral proteins, crucial for cell entry, are also the virus's Achilles' heel for they provide a target for the body's immune system and for scientists who design antiviral therapies. This chapter will describe the various agents being used to fight viral infections, including antiviral drugs, natural antiviral compounds such as the interferons, and vaccines.

## ANTIVIRAL DRUGS

A wide range of antiviral drugs have been produced to treat viral diseases for which there are no vaccines. Studies over the past 50 years have defined many of the targets that scientists can use to

## ANTIVIRAL DRUGS

| DRUG | MECHANISM OF ACTION | VIRAL TARGET |
| --- | --- | --- |
| Fuzeon *(Enfuvirtide)* | Blocks cell fusion and entry | HIV |
| Foscavir *(Foscarnet)* | Inhibits viral polymerase | Herpes, HIV |
| Viramune *(Nevirapine)* | Inhibits reverse transcriptase | HIV |
| Videx *(Didanosine)* | Inhibits reverse transcriptase | HIV |
| Valtrex *(Valacyclovir)* | Inhibits viral polymerase | Herpes, varicella |
| Zidovudine *(AZT)* | Inhibits reverse transcriptase | HIV |
| Crixivan *(Indinavir)* | Inhibits viral maturation | HIV |
| Fortovase *(Saquinavir)* | Inhibits viral maturation | HIV |
| Relenza *(Zanamivir)* | Blocks cell exit | Influenza A and B |
| Tamiflu *(Oseltamivir)* | Blocks cell exit | Influenza A and B |

Note: For each drug listed, the trade name is followed by the scientific name in italics.

treat a viral infection. Drugs are now available that can slow the progression of a viral infection by blocking the viral life cycle at the following stages: cell entry, replication of the viral genome and transcription of viral genes, assembly and maturation of daughter virions, and cell exit. Most of the antiviral drugs being used today are designed to treat infections caused by HIV, herpes, varicella, and the influenza viruses.

## Cell Entry

All viruses would be harmless microbes if they could be blocked from entering a cell. But to date, there is only one drug available that can inhibit cell fusion and entry. This drug is called Fuzeon, and it is designed to block HIV's ability to infect healthy CD4 T lymphocytes. Fuzeon must be injected with a hypodermic needle twice a day. Although moderately effective, the use of this drug is associated with many side effects. Nearly everyone who uses Fuzeon experiences injection site reactions, which range from a mild rash to itching, swelling, hardened skin or bumps. The drug can also cause serious allergic reactions, which may include the following: trouble breathing, fever with vomiting and skin rash, blood in the urine, and swelling of the feet.

## Polymerase Inhibitors

These drugs can inhibit replication and transcription of a viral genome as well as blocking reverse transcription of an RNA genome into DNA. Some of these drugs, such as Foscavir and Viramune, bind to the polymerase to inactivate it, while other drugs such as Videx, Valtrex, and Zidovudine are nucleoside analogs that block the polymerase indirectly. Incorporation of these analogs into DNA or RNA terminates growth of the daughter strand. Zidovudine, also known as azidodthymidine (AZT), was one of the first drugs of this kind to be produced. AZT is a chain terminator of DNA synthesis. This drug is an analog of thymidine in which the 3′ hydroxyl group is replaced by an azido group, consisting of three nitrogens ($N_3$). The polymerase can only add an incoming nucleoside to the hydroxyl group on the last nucleotide in the chain. If the hydroxyl group is missing, the growth of the daughter strand is terminated. Thus, in the presence of AZT the virus cannot replicate its genome.

Common side effects of AZT include nausea, headache, changes in body fat, and discoloration of fingernails and toenails. More severe side effects include anemia and bone marrow suppression, which can be treated with erythropoietin, a hormone that stimulates the production of red blood cells.

## Viral Assembly and Maturation

The HIV life cycle depends on the activity of a protease that cleaves the Gag and Gag-Pol protein precursors to form the mature virion core and to activate the reverse transcriptase. Inhibition of the protease results in noninfectious viral particles. Commonly used

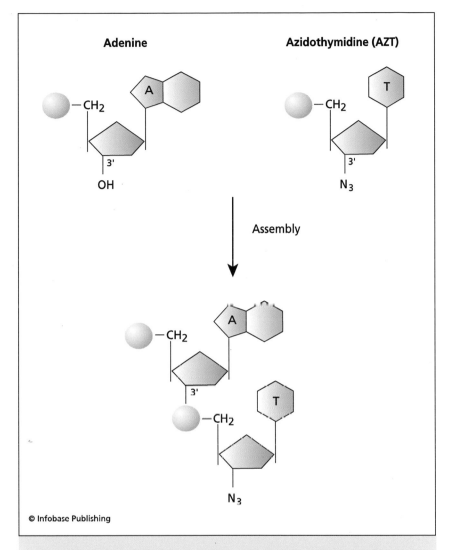

© Infobase Publishing

Azidothymidine (AZT). This compound is used to fight HIV infections. AZT is an analog of thymidine that functions as a chain terminator during DNA synthesis. Nucleotides normally have a hydroxyl group (OH) at the 3' carbon on the ribose sugar. In AZT, the 3' hydroxyl group is replaced by an azido group, consisting of three nitrogens ($N_3$). The reverse transcriptase can only add an incoming nucleoside to the hydroxyl group on the last nucleotide in the chair. If the hydroxyl group is missing, the growth of the daughter strand is terminated. Thus, in the presence of AZT the virus cannot replicate its genome.

protease inhibitors are Crixivan and Fortovase. The main side effects with Crixivan are kidney stones and metabolic abnormalities associated with elevated levels of cholesterol. Fortovase can cause nausea and abdominal discomfort.

## Cell Exit

The exit strategy employed by the influenza viruses involves the envelope glycoprotein neuraminidase. During the life cycle of these viruses, neuraminidase and hemagglutinin are deposited in the cell membrane. Contact between these viral proteins and the daughter capsid activates the neuraminidase, which in turn activates the budding stage and exit of the fully formed viruses from the cell. Two drugs, Relenza and Tamiflu, are available that can inhibit neuraminidase, thus blocking the release of the virus from the cell. Although these drugs are effective and popular, both have fairly serious side effects, which includes psychiatric problems such as abnormal or bizarre behavior, delirium, self-injury, and even suicide. Many strains of influenza viruses have become resistant to Tamiflu, but fortunately they are still susceptible to Relenza.

## INTERFERONS

Human physiology is so complex that its coordination requires three separate communication systems that operate simultaneously: the nervous system, the endocrine system, and the paracrine system. The first system, and the most familiar, is a network of neurons that make it possible for us to move our muscles, read a book, see the world, and plan our future. The endocrine system consists of special glands located throughout the body that are controlled by the pituitary gland, the master gland of the body. The pituitary gland releases hormones into the circulatory system, which control the activities of all the other glands. For example, the pituitary might release follicle-stimulating hormone (FSH) to regulate the growth and development of oocytes in the ovaries or adrenocorticotropic hormone (ACTH), which activates the release of cortisol from the adrenal gland. The paracrine system also depends upon the release of hormones, but they are not released into the blood. Instead, these hormones are intended for local communication between individual cells and are used extensively by the immune system and other cells in the body to deal with an invading microbe. The hormones are glycoproteins called cytokines.

A special group of cytokines called interferons (IFNs), because they interfere with microbial infections, are secreted by leukocytes and many other cells in response to a viral infection. The interferons not only coordinate the immune response by activating T cells and macrophages, but they also help uninfected cells resist the infection and in some cases can inhibit viral replication. Interferons are produced within hours of an infection and represent the body's first line of defense against viruses. Infected cells warn neighboring cells of a viral presence by synthesizing and releasing IFNs. The neighboring cells respond to the alarm by producing a veritable army of antiviral enzymes (encoded by interferon-stimulated genes). Three of these enzymes are protein kinase R (PKR), RNAse L, and the cell-cycle regulator P53. PKR inactivates the cell's protein translation machinery, thus reducing or completely abolishing protein synthesis. RNAse L destroys all of the cell's messenger RNA (mRNA), and P53, if it senses abnormal DNA replication, will force the cell to commit suicide. These three events block the formation of new viral particles and eliminate infected cells from the body. The interferon response is not virus-specific but is effective against any DNA or RNA virus.

About 10 distinct IFNs have been identified in mammals, seven of which are specific to humans. Human IFNs are divided into two groups based on the type of cell-surface receptor that they bind to: Type I binds to the IFN-α and IFN-β receptors, and Type II binds to the IFN-γ receptor. The IFN-α family is very large, being encoded by at least 20 different genes, whereas IFN-β and IFN-γ are each encoded by a single gene. Although focused on the same goal (i.e., responding to a viral infection), there is very little sequence homology between these three gene families. This is no doubt due to the diverse nature of the response, many parts of which are still poorly understood.

The interferon response is so rapid and so thorough that it is a wonder that viruses are able to get a foot in the door. But the battle between host and parasite is an ancient one that produces brilliant defensive and offensive strategies in a stepwise fashion. Viruses have thus managed to counter nearly every one of the cell's interferon-based defenses. Herpes viruses, for example, can block the induction of interferon-stimulated genes, and adenoviruses can block the activation of PKR.

Nevertheless, the IFN genes were cloned in the 1980s and large quantities of interferons were produced using biotechnology. Although synthetic IFN-α has been used as an effective early treatment for hepatitis B and C infections, the therapy overall has not lived up to early expectations. The main problem is that in order to be effective, the IFN dose must be very high, which produces serious side effects, some of them involving the central nervous system. In most cases, viral infections are best treated with antiviral drugs or vaccines.

## VACCINES

Edward Jenner produced the first vaccine in the late 1700s to treat smallpox. This early success was made possible to a great extent by the fact that cowpox, common in Jenner's day, produced relatively mild symptoms in humans while protecting them from smallpox. Jenner's observation that a mild form of a deadly virus could be used to produce an effective vaccine is a central strategy in vaccine production even today.

Louis Pasteur made a similar observation in 1885 when he discovered that a rabies virus isolated from rabbit nerve tissue that had been dried for 10 days could be used to immunize dogs without inducing symptoms of the disease. Such a preparation is called a live attenuated virus vaccine. Attenuation may explain why smallpox variolation, discussed previously, sometimes produced a milder disease, followed by a lasting immunity. Attenuated vaccines have the advantage of acting like the natural infection with regard to their ability to activate an immune response, without actually causing the disease (although they might cause a mild form of the disease). The alternative method for producing a vaccine is to kill the virus, usually by treating it with formaldehyde. As discussed previously, killed vaccines have many disadvantages and consequently the majority of modern vaccines are made with live attenuated viruses.

One of the most difficult aspects of producing a virus vaccine is finding a way to grow the virus either in a test animal or a cell culturing system. Once this is achieved, scientists can determine the best way to attenuate the virus and can then grow enough of

## VIRUS VACCINES

| VACCINE | TYPE | CELL CULTURE |
|---|---|---|
| Hepatitis | Killed | Human fibroblasts |
| Influenza A and B | Killed | Chicken eggs |
| Influenza A and B | Live | Chicken eggs |
| Measles | Live | Chicken embryo fibroblasts |
| Mumps | Live | Chicken embryo fibroblasts |
| Polio | Live | Monkey kidney cells |
| Rabies | Killed | Chicken embryo fibroblasts |
| Rubella | Live | Human fibroblasts |
| Varicella | Live | Human fibroblasts |
| Yellow fever | Live | Chicken eggs |
| Zoster | Live | Human fibroblasts |

it for large-scale vaccine production. Pasteur grew the rabies virus in rabbits, an animal that normally contracts the disease, and then prepared the vaccine from the nervous tissue. But virologists wanted a cell culturing system similar to what Pasteur and Koch had developed for culturing bacteria. These cultures consisted of nutrient agar in a glass dish, fitted with a lid to keep it sterile (called a petri dish or culture plate). All they had to do was spread a bit of the bacterial sample on the plate and they could grow as much of it as they needed.

But viruses are parasites that only grow inside cells. So the problem became one of finding ways to culture a variety of animal cells and tissues and then testing each of the cultures to see if they would support the growth of specific viruses. The basic problem was solved in the late 1920s by the British bacteriologists Hugh and Mary Maitland, who succeeded in growing vaccinia viruses in petri dishes or a tissue culture suspension (a small tube or flask containing a nutrient broth and the cultured cells. The container was rolled or agitated to keep the cells in suspension). During this same period, Max Theiler, an American virologist studying yellow fever, discovered a simple way of attenuating the virus by culturing it through several passages in a tissue culture derived from chicken embryos (i.e., transferring the infected cells to new dishes a dozen times or more). The exact mechanism by which the virus is attenuated is unknown. Presumably, the extended period in cell culture induces a number of genetic mutations, which convert the virus to a less virulent strain. In this way, Theiler produced the 17D strain of the yellow fever virus, from which he made a very safe and effective vaccine.

Throughout the first half of the 1900s, virologists developed tissue culture systems containing a great variety of animal and human tissues. They tested cultures derived from kidney epithelial cells, human amniotic fluid (for suspension cultures), and nerve, muscle, and skin cells derived from chicken, mouse, and monkey embryos. The American virologist, John F. Enders, tested many of these culture systems for their ability to support the growth of a

Production of an influenza vaccine. A microbiologist "candles" a chicken egg to determine the viability of the chicken embryo. *(U.S. Centers for Disease Control/Laura R. Zambuto)*

variety of viruses, including polio and measles viruses. He was also one of the first to grow viruses in fertile chicken eggs, used today for producing the seasonal flu vaccine. Enders and Thomas H. Weller and Frederick C. Robbins, two of his colleagues, won the Nobel Prize in physiology or medicine in 1954 for establishing what are now the basic procedures for culturing viruses. Albert Sabin used these procedures to produce the first attenuated polio vaccine in 1960, and Enders went on to produce an effective measles vaccine the following year.

## IMMUNIZATION POLICIES

In the United States, the federal government plays an important role in immunization programs. Although vaccines are made by private companies and immunization policies are set by each state, the Department of Health and Human Services (HHS) regulates vaccine production. In addition, HHS is responsible for making vaccines available to the states and for ensuring an adequate supply.

Regulating vaccination of schoolchildren is an important aspect of the federal mandate. School immunization laws were established to control smallpox and have subsequently been used to avoid epidemics of measles, whooping cough, and polio. Currently, all 50 states have school immunization laws, although the exact requirements differ from state to state.

All 50 states allow vaccination exemptions for medical reasons; 48 states allow exemptions for religious reasons; and 20 states allow exemptions for philosophical reasons. Medical exemptions are determined by a physician and are recommended for children who are allergic to some component of the vaccine or have a weakened immune system. In most states, a child can attend school or day care if they obtain an exemption, but they are expected to remain at home if there is an outbreak of vaccine-preventable disease. Mississippi and West Virginia are the only states that do not allow exemptions for religious reasons. The 20 states that allow exemptions for philosophical reasons are Arizona, Arkansas, Idaho, Louisiana, Maine, Michigan, Minnesota, Missouri, Nebraska, New Mexico, North Dakota, California, Colorado, Ohio, Oklahoma, Texas, Utah, Vermont, Washington, and Wisconsin.

Although adults are certainly within their rights to refuse vaccination, the question of whether they also have the right to prevent their children from being vaccinated is debatable; especially when the vaccine targets a crippling and often deadly disease such as polio. The CDC has recently estimated that nearly 40,000 American children have not been vaccinated because their parents object to it on religious or philosophical grounds. Estimates have shown that these children are 10 to 35 times more likely to develop a viral disease. There is no legal penalty for parents who obtain exemptions for religious or philosophical reasons, but in some states they can be held civilly liable if their child were to transmit a disease to another child.

## SUMMARY

Viral infections are fought with antiviral drugs, synthetic interferons, and vaccines. Antiviral drugs target every aspect of the viral life cycle, beginning with binding to the cell membrane, replication and transcription of viral genes, assembly of the daughter virions, and the final exit from the cell. These drugs have proven to be effective, particularly in controlling HIV, but they usually produce serious side effects. Interferons are part of the body's first line of defense against viral infections. They function reasonably well and are often able to blunt the viral attack, giving the immune system time to produce the necessary antibodies. Synthetic interferons have been disappointing, mainly because the high doses required generally produce serious side effects. By far the most effective way to treat a viral infection is with a vaccine. Vaccine production is a very sophisticated business that depends on a deep knowledge of viruses and tissue culturing techniques. Vaccines are currently available to treat and prevent most viral infections. Notable exceptions are HIV and the hemorrhagic fever viruses. The U.S. government regulates the production and distribution of vaccines. Although vaccinations for common viral diseases are generally required for schoolchildren, there are provisions in the laws that allow exemptions for medical, religious, and philosophical reasons.

# 10

# Resource Center

Virology depends on an understanding of the cell and the experimental procedures, known as biotechnology, that are used to explore cell structure and function. This chapter discusses the basics of cell biology, biotechnology, and some of the methods used in virology. Additional material is included that covers the following relevant topics: the design of clinical trials, gene and protein nomenclature, and a table of weights and measures.

## CELL BIOLOGY

A cell is a microscopic life-form made from a variety of nature's building blocks. The smallest of these building blocks are subatomic particles known as quarks and leptons that form protons, neutrons, and electrons, which in turn form atoms. Scientists have identified more than 200 atoms, each of which represents a fundamental element of nature; carbon, oxygen, and nitrogen are common

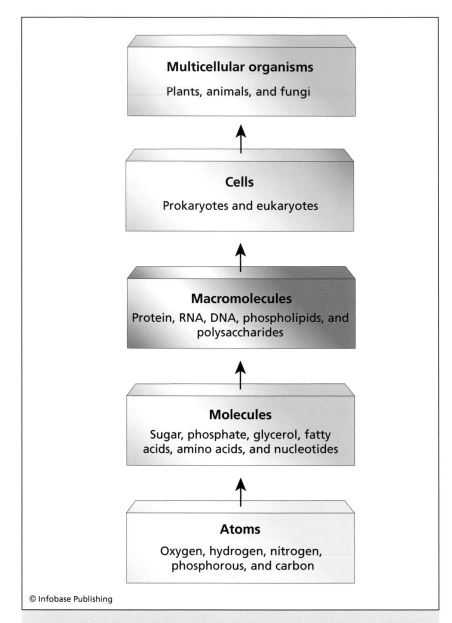

Nature's building blocks. Particles known as quarks and leptons, created in the heat of the big bang, formed the first atoms, which combined to form molecules in the oceans of the young Earth. Heat and electrical storms promoted the formation of macromolecules, providing the building blocks for cells, which in turn went on to form multicellular organisms.

examples. Atoms, in their turn, can associate with one another to form another kind of building block known as a molecule. Sugar, for example, is a molecule constructed from carbon, oxygen, and hydrogen, while ordinary table salt is a molecule consisting of two elements, sodium and chloride. Molecules can link up with one another to form yet another kind of building block known as a macromolecule. Macromolecules, present in the atmosphere of the young Earth, gave rise to cells, which in turn went on to form multicellular organisms; in forming those organisms, cells became a new kind of building block.

## The Origin of Life

Molecules essential for life are thought to have formed spontaneously in the oceans of the primordial Earth about 4 billion years ago. Under the influence of a hot stormy environment, the molecules combined to produce macromolecules, which in turn formed microscopic bubbles that were bounded by a sturdy macromolecular membrane analogous to the skin on a grape. It took about half a billion years for the prebiotic bubbles to evolve into the first cells, known as prokaryotes, and another 1 billion years for those cells to evolve into the eukaryotes. Prokaryotes, also known as bacteria, are small cells (about five micrometers in diameter) that have a relatively simple structure and a genome consisting of about 4,000 genes. Eukaryotes are much larger (about 30 micrometers in diameter), with a complex internal structure and a very large genome, often exceeding 20,000 genes. These genes are kept in a special organelle called the nucleus (eukaryote means "true nucleus"). Prokaryotes are all single-cell organisms, although some can form short chains or temporary fruiting bodies. Eukaryotes, on the other hand, gave rise to all of the multicellular plants and animals that now inhabit the Earth.

## A Typical Eukaryote

Eukaryotes assume a variety of shapes that are variations on the simple spheres from which they originated. Viewed from the side,

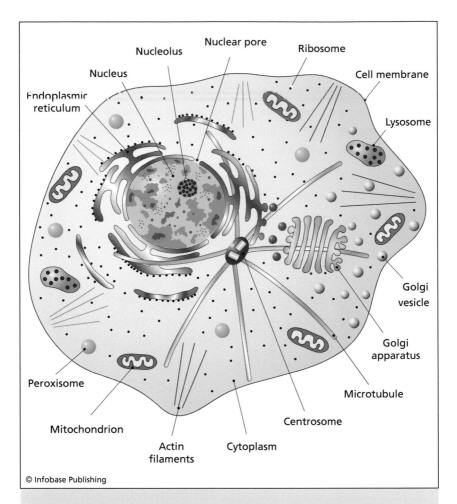

Nuclear pore

Nucleolus

Ribosome

Nucleus

Cell membrane

Endoplasmic
reticulum

Lysosome

Golgi
vesicle

Golgi
apparatus

Microtubule

Peroxisome

Centrosome

Mitochondrion

Actin
filaments

Cytoplasm

The eukaryote cell. The structural components shown here are present in organisms as diverse as protozoans, plants, and animals. The
nucleus contains the DNA genome and an assembly plant for ribosomal subunits (the nucleolus). The endoplasmic reticulum (ER) and
the Golgi work together to modify proteins, most of which are destined for the cell membrane. These proteins travel from the ER to the
Golgi and from the Golgi to their final destination in transport vesicles
(red and yellow spheres). Mitochondria provide the cell with energy
in the form of ATP. Ribosomes, some of which are attached to the ER,
synthesize proteins. Lysosomes and peroxisomes recycle cellular material. The microtubules and centrosome form the spindle apparatus
for moving chromosomes to the daughter cells during cell division.
Actin and other protein filaments form a weblike cytoskeleton.

they often have a galactic profile, with a central bulge (the nucleus), tapering to a thin perimeter. The internal structure is complex, being dominated by a large number of organelles.

The functional organization of a eukaryote is analogous to a carpentry shop, which is usually divided into two main areas: the shop floor where the machinery, building materials, and finishing rooms are kept, and the shop office, where the work is coordinated and where the blueprints are stored for everything the shop makes. Carpentry shops keep a blueprint on file for every item that is made. When the shop receives an order, perhaps for a chair, someone in the office makes a copy of the chair's blueprint and delivers it to the carpenters on the shop floor. In this way the master copy is kept out of harm's way, safely stored in the filing cabinet. The carpenters, using the blueprint copy and the materials and tools at hand, build the chair, and then they send it into a special room where it is painted. After the chair is painted, it is taken to another room where it is polished and then packaged for delivery. The energy for all of this activity comes through the electrical wires, which are connected to a power generator somewhere in the local vicinity. The shop communicates with other shops and its customers by using the telephone, e-mail, or postal service.

In the cell, the shop floor is called the cytoplasm and the shop office is the nucleus. Eukaryotes make a large number of proteins, and they keep a blueprint for each one, only in this case the blueprints are not pictures on pieces of paper but molecules of deoxyribonucleic acid (DNA) that are kept in the nucleus. A cellular blueprint is called a gene, and a typical cell has thousands of them. A human cell, for example, has 30,000 genes, all of which are kept on 46 separate DNA molecules known as chromosomes (23 from each parent). When the cell decides to make a protein, it begins by making a ribonucleic acid (RNA) copy of the protein's gene. This blueprint copy, known as messenger RNA, is made in the nucleus and delivered to the cell's carpenters in the cytoplasm. These carpenters are enzymes that control and regulate all of the cell's chemical reactions. Some of the enzymes are part of a complex protein-synthesizing machine known as a ribosome. Cytoplasmic enzymes and the ribosomes synthesize proteins using mRNA as the template, after which many of the proteins are sent to a compartment, known as the endoplasmic reticulum (ER), where they are glycosylated or "painted" with sugar molecules. From there they are shipped to another compartment called the Golgi apparatus, where the glycosylation is refined before the finished products, now looking like molecular trees, are loaded into transport bubbles and shipped to their final destination.

The shape of the cell is maintained by an internal cytoskeleton comprising actin and intermediate filaments. Mitochondria, once free-living prokaryotes, provide the cell with energy in the form of adenosine triphosphate (ATP). The production of ATP is carried out by an assembly of metal-containing proteins, called the electron transport chain, located in the mitochondrion inner membrane. Lysosomes and peroxisomes process and recycle cellular material and molecules. The cell communicates with other cells and the outside world through the glycocalyx, an enormous collection of glycoproteins and glycolipids that covers the cell surface. Producing and maintaining the glycocalyx is the principal function of the ER and Golgi apparatus and a major priority for all eukaryotes.

Cells are biochemical entities that synthesize many thousands of molecules. Studying these chemicals, as well as the biochemistry of the cell, would be extremely difficult were it not for the fact that most of the chemical variation is based on six types of molecules that are assembled into just five types of macromolecules. The six basic molecules are: amino acids, phosphate, glycerol, sugars, fatty acids, and nucleotides. The five macromolecules are: proteins, DNA, RNA, phospholipids, and sugar polymers called polysaccharides.

## Molecules of the Cell

Amino acids have a simple core structure consisting of an amino group, a carboxyl group, and a variable R group attached to a carbon atom. There are 20 different kinds of amino acids, each with a

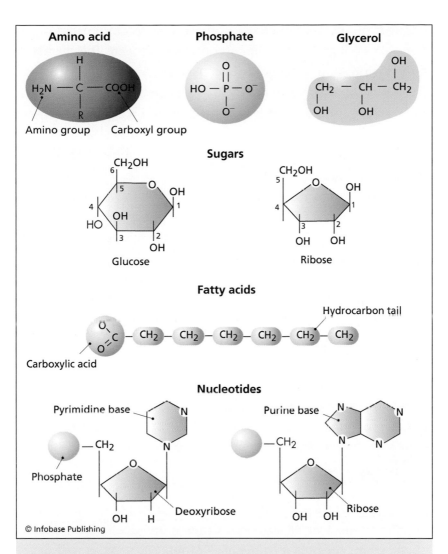

Molecules of the cell. Amino acids are the building blocks for proteins. Phosphate is an important component of many other molecules and is added to proteins to modify their behavior. Glycerol is an alcohol that is an important ingredient in cell membranes and fat. Sugars, like glucose, are a primary energy source for most cells and also have many structural functions. Fatty acids are involved in the production of cell membranes and storage of fat. Nucleotides are the building blocks for DNA and RNA. Note that the sugar carbon atoms are numbered. P: Phosphate C: Carbon H: Hydrogen O: Oxygen N: Nitrogen R: Variable molecular group

unique R group. The simplest and most ancient amino acid is glycine, with an R group that consists only of hydrogen. The chemistry of the various amino acids varies considerably: Some carry a positive electric charge, while others are negatively charged or electrically neutral; some are water soluble (hydrophilic), while others are hydrophobic.

Phosphates are extremely important molecules that are used in the construction, or modification, of many other molecules. They are also used to store chemical-bond energy in the form of adenosine triphosphate (ATP). The production of phosphate-to-phosphate chemical bonds for use as an energy source is an ancient cellular process, dating back at least 2 billion years.

Glycerol is a simple three-carbon alcohol that is an important component of cell membranes and fat reservoirs. This molecule may have stabilized the membranes of prebiotic bubbles. Interestingly, it is often used today as an ingredient in a solution for making long-lasting soap bubbles.

Sugars are versatile molecules, belonging to a general class of compounds known as carbohydrates that serve a structural role as well as providing energy for the cell. Glucose, a six-carbon sugar, is the primary energy source for most cells and the principal sugar used to glycosylate the proteins and lipids that form the outer coat of all cells. Plants have exploited the structural potential of sugars in their production of cellulose; wood, bark, grasses, and reeds are all polymers of glucose and other monosaccharides. Ribose, a five-carbon sugar, is a component of nucleic acids as well as the cell's main energy depot, ATP. Ribose carbons are numbered as 1' (1 prime), 2', and so on. Consequently, references to nucleic acids, which include ribose, often refer to the 3' or 5' carbon.

Fatty acids consist of a carboxyl group (the hydrated form is called carboxylic acid) linked to a hydrophobic hydrocarbon tail. These molecules are used in the construction of cell membranes and fat. The hydrophobic nature of fatty acids is critically important to the normal function of the cell membrane since it prevents the passive entry of water and water-soluble molecules.

Nucleotides are building blocks for DNA and RNA. These molecules consist of three components: a phosphate, a ribose sugar, and a nitrogenous (nitrogen-containing) ring compound that behaves as a base in solution (a base is a substance that can accept a proton in solution). Nucleotide bases appear in two forms: a single-ring nitrogenous base, called a pyrimidine, and a double-ringed base, called a purine. There are two kinds of purines (adenine and guanine) and three pyrimidines (uracil, cytosine, and thymine). Uracil is specific to RNA, substituting for thymine. In addition, RNA nucleotides contain ribose, whereas DNA nucleotides contain deoxyribose (hence the origin of their names). Ribose has a hydroxyl (OH) group attached to both the 2′ and 3′ carbons, whereas deoxyribose is missing the 2′ hydroxyl group.

## Macromolecules of the Cell

The six basic molecules are used by all cells to construct five essential macromolecules: proteins, RNA, DNA, phospholipids, and polysaccharides. Macromolecules have primary, secondary, and tertiary structural levels. The primary structural level refers to the chain that is formed by linking the building blocks together. The secondary structure involves the bending of the linear chain to form a three-dimensional object. Tertiary structural elements involve the formation of chemical bonds between some of the building blocks in the chain to stabilize the secondary structure. A quaternary structure can also occur when two identical molecules interact to form a dimer or double molecule.

Proteins are long chains or polymers of amino acids. The primary structure is held together by peptide bonds that link the carboxyl end of one amino acid to the amino end of a second amino acid. Thus once constructed, every protein has an amino end and a carboxyl end. An average protein consists of about 400

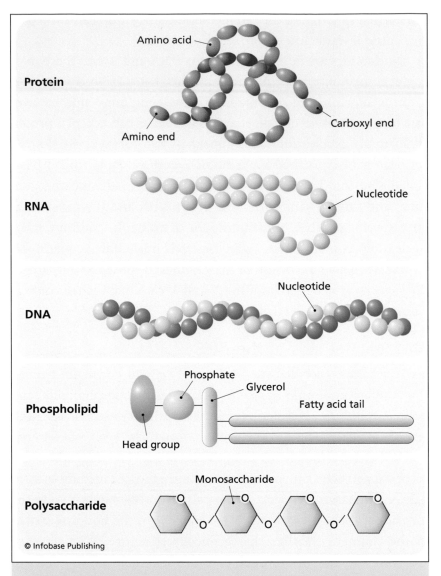

Macromolecules of the cell. Protein is made from amino acids linked together to form a long chain that can fold up into a three-dimensional structure. RNA and DNA are long chains of nucleotides. RNA is generally single-stranded but can form localized double-stranded regions. DNA is a double-stranded helix, with one strand coiling around the other. A phospholipid is composed of a hydrophilic head-group, a phosphate, a glycerol molecule, and two hydrophobic fatty acid tails. Polysaccharides are sugar polymers.

amino acids. There are 21 naturally occurring amino acids; with this number the cell can produce an almost infinite variety of proteins. Evolution and natural selection, however, have weeded out most of these, so that eukaryote cells function well with 10,000 to 30,000 different proteins. In addition, this select group of proteins has been conserved over the past 2 billion years (i.e., most of the proteins found in yeast can also be found, in modified form, in humans and other higher organisms). The secondary structure of a protein depends on the amino acid sequence and can be quite complicated, often producing three-dimensional structures possessing multiple functions.

RNA is a polymer of the ribonucleotides adenine, uracil, cytosine, and guanine. RNA is generally single-stranded, but it can form localized double-stranded regions by a process known as complementary base pairing, whereby adenine forms a bond with uracil and cytosine pairs with guanine. RNA is involved in the synthesis of proteins and is a structural and enzymatic component of ribosomes.

DNA is a double-stranded nucleic acid. This macromolecule encodes cellular genes and is constructed from adenine, thymine, cytosine, and guanine deoxyribonucleotides. The two DNA strands coil around each other like strands in a piece of rope, creating a double helix. The two strands are complementary throughout the length of the molecule: adenine pairs with thymine, and cytosine pairs with guanine. Thus if the sequence of one strand is known to be ATCGTC, the sequence of the other strand must be TAGCAG.

Phospholipids are the main component in cell membranes; these macromolecules are composed of a polar head group (usually an alcohol), a phosphate, glycerol, and two hydrophobic fatty acid tails. Fat that is stored in the body as an energy reserve has a structure similar to a phospholipid, being composed of three fatty acid chains attached to a molecule of glycerol. The third fatty acid takes the place of the phosphate and head group of a phospholipid.

Polysaccharides are sugar polymers consisting of two or more monosaccharides. Disaccharides (two monosaccharides) and oligosaccharides (about three to 12 monosaccharides) are attached to proteins and lipids destined for the cell surface or the extracellular matrix. Polysaccharides, such as glycogen and starch, may contain several hundred monosaccharides, and are stored in cells as an energy reserve.

## Basic Cellular Functions

There are six basic cellular functions: DNA replication, DNA maintenance, gene expression, power generation, cell division, and cell communication. DNA replication usually occurs in conjunction with cell division, but there are exceptions known as polyploidization (see the Glossary). Gene expression refers to the process whereby the information stored in a gene is used to synthesize RNA or protein. The production of power is accomplished by extracting energy from food molecules and then storing that energy in a form that is readily available to the cell. Cells communicate with their environment and with other cells. The communication hardware consists of a variety of special macromolecules that are embedded in the cell membrane.

## *DNA Replication*

Replication is made possible by the complementarity of the two DNA strands. Since adenine (A) always pairs with thymine (T) and guanine (G) always pairs with cytosine (C), replication enzymes are able to duplicate the molecule by treating each of the original strands as templates for the new strands. For example, if a portion of the template strand reads ATCGTTGC, the new strand will be TAGCAACG.

DNA replication requires the coordinated effort of a team of enzymes, led by DNA helicase and primase. The helicase separates the two DNA strands at the astonishing rate of 1,000 nucleotides every second. This enzyme gets its name from the fact that it unwinds the DNA helix as it separates the two strands. The enzyme that is directly responsible for reading the template strand and for synthesizing the new daughter strand is called DNA polymerase. This enzyme also has an editorial function; it checks the preceding nucleotide to make sure it is correct before it adds a nucleotide to the growing chain. The editor function of this enzyme introduces an interesting problem. How can the polymerase add the very first nucleotide, when it has to check a preceding nucleotide before adding a new one? A special enzyme, called primase, which is attached to the helicase, solves this problem. Primase synthesizes short pieces of RNA that form a DNA-RNA double-stranded region. The RNA becomes a temporary part of the daughter strand, thus priming the DNA polymerase by providing the crucial first nucleotide in the new strand. Once the chromosome is duplicated, DNA repair enzymes, discussed below, remove the RNA primers and replace them with DNA nucleotides.

## DNA Maintenance

Every day in a typical human cell, thousands of nucleotides are being damaged by spontaneous chemical events, environmental pollutants, and radiation. In many cases, it takes only a single defective nucleotide within the coding region of a gene to produce an inactive, mutant protein. The most common forms of DNA damage are depurination and deamination. Depurination is the loss of a purine base (guanine or adenine), resulting in a gap in the DNA sequence, referred to as a "missing tooth." Deamination converts cytosine to uracil, a base that is normally found only in RNA.

About 5,000 purines are lost from each human cell every day, and over the same time period 100 cytosines are deaminated per cell. Depurination and deamination produce a great deal of damage, and in either case the daughter strand ends up with a missing nucleotide, and possibly a mutated gene, as the DNA replication machinery simply bypasses the uracil or the missing tooth. If left unrepaired, the mutated genes will be passed on to all daughter cells, with catastrophic consequences for the organism as a whole.

DNA damage caused by depurination is repaired by special nuclear proteins that detect the missing tooth, excise about 10 nucleotides on either side of the damage, and then, using the complementary strand as a guide, reconstruct the strand correctly. Deamination is dealt with by a special group of DNA repair enzymes known as base-flippers. These enzymes inspect the DNA one nucleotide at a time. After binding to a nucleotide, a base-flipper breaks the hydrogen bonds holding the nucleotide to its complementary partner. It then performs the maneuver for which it gets its name. Holding onto the nucleotide, it rotates the base a full 180 degrees, inspects it carefully, and, if it detects any damage, cuts the base out and discards it. In this case the base-flipper leaves the final repair to the missing-tooth crew that detects and repairs the gap as described previously. If the nucleotide is normal, the base-flipper rotates it back into place and reseals the hydrogen bonds. Scientists have estimated that these maintenance crews inspect and repair the entire genome of a typical human cell in less than 24 hours.

## Gene Expression

Genes encode proteins and several kinds of RNA. Extracting the coded information from DNA requires two sequential processes known as transcription and translation. A gene is said to be expressed when either or both of these processes have been completed. Transcription, catalyzed by the enzyme RNA polymerase, copies one strand of the DNA into a complementary strand of mRNA, which is sent to the cytoplasm, where it joins with a ribosome. Translation is a process that is orchestrated by the ribosomes. These particles synthesize proteins using mRNA and the genetic code as guides. The ribosome can synthesize any protein specified by the mRNA, and the mRNA can be translated many times before it is recycled.

Some RNAs, such as ribosomal RNA and transfer RNA, are never translated. Ribosomal RNA (rRNA) is a structural and enzymatic component of ribosomes. Transfer RNA (tRNA), though separate from the ribosome, is part of the translation machinery.

The genetic code provides a way for the translation machinery to interpret the sequence information stored in the DNA molecule and represented by mRNA. DNA is a linear sequence of four different kinds of nucleotides, so the simplest code could be one in which each nucleotide specifies a different amino acid; that is, adenine coding for the amino acid glycine, cytosine for lysine, and so on. The earliest cells may have used this coding system, but it is limited to the construction of proteins consisting of only four different kinds of amino acids. Eventually a more elaborate code evolved in which a combination of three out of the four possible DNA nucleotides, called codons, specifies a single amino acid. With this scheme it is possible to have a unique code for each of the 20 naturally occurring amino acids. For example, the codon AGC specifies the amino acid serine, whereas TGC specifies the amino acid cysteine. Thus, a gene may be viewed as a long continuous sequence of codons. However, not all codons specify an amino acid. The sequence TGA signals the end of the gene, and a special codon, ATG, signals the start site, in addition to specifying the amino acid methionine. Consequently, all proteins begin with this amino acid, although it is sometimes removed once construction of the protein is complete. As mentioned above, an average protein may consist of 300 to 400 amino acids; since the codon consists of three nucleotides for each amino acid, a typical gene may be 900 to 1,200 nucleotides long.

## Power Generation

Dietary fats, sugars, and proteins, not targeted for growth, storage, or repairs, are converted to ATP by the mitochondria. This process requires a number of metal-binding proteins, called the respiratory chain (also known as the electron transport chain), and a special ion channel-enzyme called ATP synthase. The respiratory chain consists of three major components: NADH dehydrogenase, cytochrome b, and cytochrome oxidase. All of these components are protein complexes with an iron (NADH dehydrogenase, cytochrome b) or a copper core (cytochrome oxidase), and together with the ATP synthase are located in the inner membrane of the mitochondria.

The respiratory chain is analogous to an electric cable that transports electricity from a hydroelectric dam to our homes, where it is used to turn on lights or to run stereos. The human body, like that of all animals, generates electricity by processing food molecules through a metabolic pathway called the Krebs cycle, also located within the mitochondria. The electrons (electricity) so generated are transferred to hydrogen ions, which quickly bind to a special nucleotide called nicotinamide adenine dinucleotide (NAD). Binding of the hydrogen ion to NAD is noted by abbreviating the resulting molecule as NADH. The electrons begin their journey down the respiratory chain when NADH binds to NADH dehydrogenase, the first component in the chain. This enzyme does just what its name implies: It removes the hydrogen from NADH, releasing the stored electrons, which are conducted through the chain by the iron and copper as though they were traveling along an electric wire. As the electrons travel from one end of the chain to the other, they energize the synthesis of ATP, which is released from the mitochondria for use by the cell. All electrical circuits must have a ground, that is, the electrons need someplace to go once they have completed the circuit. In the case of the respiratory chain, the ground is oxygen. After passing through cytochrome oxidase, the last component in the chain, the electrons are picked up by oxygen, which combines with hydrogen ions to form water.

## The Cell Cycle

Free-living single cells divide as a way of reproducing their kind. Among plants and animals, cells divide as the organism grows from a seed or an embryo into a mature individual. This form of cell division, in which the parent cell divides into two identical daughter cells, is called mitosis. A second form of cell division, known as meiosis, is intended for sexual reproduction and occurs exclusively in gonads.

Cell division is part of a grander process known as the cell cycle, which consists of two phases: interphase and M phase (meiosis or mitosis). Interphase is divided into three subphases called Gap 1 ($G_1$), S phase (a period of DNA synthesis), and Gap 2 ($G_2$). The conclusion of interphase, and with it the termination of $G_2$, occurs with division of the cell and a return to $G_1$. Cells may leave the cycle by entering a special phase called $G_0$. Some cells, such as postmitotic neurons in an animal's brain, remain in $G_0$ for the life of the organism. For most cells, the completion of the cycle, known as the generation time, can take 30 to 60 minutes.

Cells grow continuously during interphase while preparing for the next round of division. Two notable events are the duplication of the spindle (the centrosome and associated microtubules), a structure that is crucial for the movement of the chromosomes during cell division, and the appearance of an enzyme called maturation-promoting factor (MPF) at the end of $G_2$. MPF phosphorylates histones, proteins that bind to the DNA, and when phosphorylated compact (or condense) the chromosomes in preparation for cell division. MPF is also responsible for the breakdown of the nuclear membrane. When cell division is complete, MPF disappears, allowing the chromosomes to decondense and the nuclear envelope to re-form. Completion of a normal cell cycle always involves the division of a cell into two daughter cells, either meiotically or mitotically.

Cell division is such a complex process that many things can, and do, go wrong. Cell cycle monitors, consisting of a team of enzymes, check to make sure that everything is going well each time a cell divides, and if it is not, those monitors stop the cell from dividing until the problem is corrected. If the damage cannot be repaired, a cell remains stuck in midstream for the remainder of

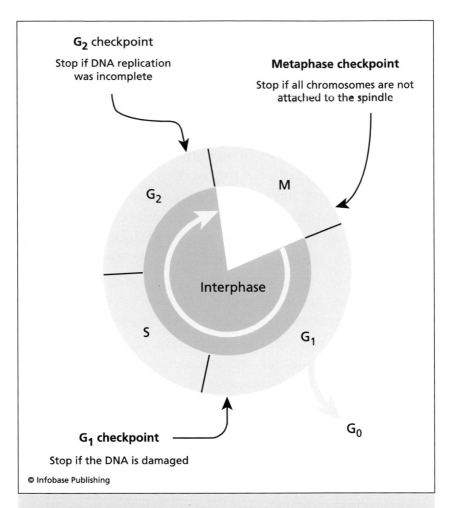

**G$_2$ checkpoint**

Stop if DNA replication
was incomplete

**Metaphase checkpoint**

Stop if all chromosomes are not
attached to the spindle

M

G$_2$

Interphase

S

G$_1$

**G$_1$ checkpoint**

Stop if the DNA is damaged

G$_0$

© Infobase Publishing

The cell cycle. Many cells spend their time cycling between interphase and M phase (cell division by mitosis or meiosis). Interphase
is divided into three subphases: Gap 1(G1), S phase (DNA synthesis),
and Gap 2 (G2). Cells may exit the cycle by entering G0. The cell cycle
is equipped with three checkpoints to ensure the daughter cells are
identical and there is no genetic damage. The yellow arrow indicates
the direction of the cycle.

its life. If this happens to a cell in an animal's body, it is forced to
commit suicide, in a process called apoptosis, by other cells in the
immediate neighborhood or by the immune system.

The cell cycle includes three checkpoints: The first is a DNA damage checkpoint that occurs in $G_1$. The monitors check for damage that may have occurred as a result of the last cell cycle or were caused by something in the environment, such as UV radiation or toxic chemicals. If damage is detected, DNA synthesis is blocked until it can be repaired. The second checkpoint occurs in $G_2$, where the monitors make sure errors were not introduced when the chromosomes were duplicated during S-phase. The $G_1$ and $G_2$ checkpoints are sometimes referred to collectively as DNA damage checkpoints. The third and final checkpoint occurs in M-phase, to ensure that all of the chromosomes are properly attached to the spindle. This checkpoint is intended to prevent gross abnormalities in the daughter cells with regard to chromosome number. If a chromosome fails to attach to the spindle, one daughter cell will end up with too many chromosomes, while the other will have too few.

### Mitosis

Mitosis is divided into four stages: prophase, metaphase, anaphase, and telophase. The behavior and movement of the chromosomes characterize each stage. At prophase, DNA replication has already occurred and the nuclear membrane begins to break down. Condensation of the duplicated chromosomes initiates the phase (i.e., the very long, thin chromosomes are folded up to produce short, thick chromosomes that are easy to move and maneuver). Under the microscope the chromosomes become visible as X-shaped structures, which are the two duplicated chromosomes, often called sister chromatids. A special region of each chromosome, called a centromere, holds the chromatids together. Proteins bind to the centromere to form a structure called the kinetochore. The centrosome is duplicated, and the two migrate to opposite ends of the cell.

During metaphase the chromosomes are sorted out and aligned between the two centrosomes. By this time the nuclear membrane has completely broken down. The two centrosomes and the microtubules fanning out between them form the mitotic spindle. The area in between the spindles, where the chromosomes are aligned, is known as the metaphase plate. Some of the microtubules make contact with the kinetochores, while others overlap, with motor proteins situated in between.

Anaphase begins when the duplicated chromosomes move to opposite poles of the cell. The first step is the release of an enzyme that breaks the bonds holding the kinetochores together, thus allowing the sister chromatids to separate from each other while remaining bound to their respective microtubules. Motor proteins, using energy supplied by ATP, move along the microtubule dragging the chromosomes to opposite ends of the cell.

During telophase the daughter chromosomes arrive at the spindle poles and decondense to form the relaxed chromosomes characteristic of interphase nuclei. The nuclear envelope begins forming around the chromosomes, marking the end of mitosis. By the end of telophase individual chromosomes are no longer distinguishable and are referred to as chromatin. While the nuclear membrane reforms, a contractile ring, made of the proteins myosin and actin, begins pinching the parental cell in two. This stage, separate from mitosis, is called cytokinesis and leads to the formation of two daughter cells, each with one nucleus.

### Meiosis

Many eukaryotes reproduce sexually through the fusion of gametes (eggs and sperm). If gametes were produced mitotically, a catastrophic growth in the number of chromosomes would occur each time a sperm fertilized an egg. Meiosis is a special form of cell division that prevents this from happening by producing haploid gametes, each possessing half as many chromosomes as the diploid cell. When haploid gametes fuse, they produce an embryo with the correct number of chromosomes

Unlike mitosis, which produces two identical daughter cells, meiosis produces four genetically unique daughter cells that have half the number of chromosomes found in the parent cell. This is possible because meiosis consists of two rounds of cell division, called meiosis I and meiosis II, with only one round of DNA synthesis. Microbiologists discovered meiosis almost 100 years ago by comparing the number of chromosomes in somatic cells and germ cells. The roundworm, for example, was found to have four chromosomes in its somatic cells, but only two in its gametes. Many other studies also compared the amount of DNA in nuclei from somatic cells and gonads, always with the same result: The amount of DNA in somatic cells is at least double the amount in fully mature gametes.

Meiotic divisions are divided into the four mitotic stages discussed above. Indeed, meiosis II is virtually identical to a mitotic division. Meiosis I resembles mitosis, but close examination shows two important differences: Gene swapping occurs between homologous chromosomes in prophase, producing recombinant chromosomes, and the distribution of maternal and paternal chromosomes to different daughter cells. At the end of meiosis I, one of the daughter cells contains a mixture of normal and recombinant maternal chromosomes, and the other contains normal and recombinant paternal chromosomes. During meiosis II, the duplicated chromosomes are distributed to different daughter cells, yielding four, genetically unique cells: paternal, paternal recombinant, maternal, and maternal recombinant. Mixing genetic material in this way is unique to meiosis, and it is one of the reasons sexual reproduction has been such a powerful evolutionary force.

## Cell Communication

A forest of glycoproteins and glycolipids covers the surface of every cell like trees on the surface of the Earth. The cell's forest is called the glycocalyx, and many of its trees function like sensory antennae. Cells use these antennae to communicate with their environment and with other cells. In multicellular organisms, the glycocalyx also plays an important role in holding cells together. In this case, the antennae of adjacent cells are connected to one another through the formation of chemical bonds.

The sensory antennae, also known as receptors, are linked to a variety of secondary molecules that serve to relay messages to the interior of the cell. These molecules, some of which are called second messengers, may activate machinery in the cytoplasm, or they may enter the nucleus to activate gene expression. The signals that a cell receives are of many different kinds but generally fall into one of five categories: 1) proliferation, which stimulates the cell to grow and divide; 2) activation, which is a request for the cell to synthesize and release specific molecules; 3) deactivation, which serves as a brake for a previous activation signal; 4) navigation, which helps direct the cell to a specific location. This is very important for free-living cells hunting for food and for immune system cells that are hunting for invading microorganisms; 5) termination, which is a signal that orders the cell to commit suicide. This death signal occurs during embryonic development (e.g., the loss of webbing between the fingers and toes) and during an infection. In some cases, the only way the immune system can deal with an invading pathogenic microbe is to order some of the infected cells to commit suicide. This process is known as apoptosis.

## BIOTECHNOLOGY

Biotechnology (also known as recombinant DNA technology) consists of several procedures that are used to study the structure and function of genes and their products. Central to this technology is the ability to clone specific pieces of DNA and to construct libraries of these DNA fragments that represent the genetic repertoire of an entire organism or a specific cell type. With these libraries at hand, scientists have been able to study the cell and whole organisms in unprecedented detail. The information so gained has revolutionized biology as well as many other disciplines, including medical science, pharmacology, psychiatry, and anthropology, to name but a few.

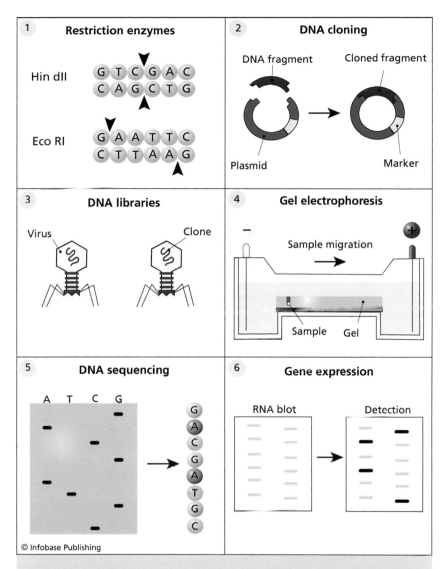

**1** Restriction enzymes

Hin dII

Eco RI

**2** DNA cloning

DNA fragment

Cloned fragment

Plasmid

Marker

**3** DNA libraries

Virus

Clone

**4** Gel electrophoresis

Sample migration

Sample    Gel

**5** DNA sequencing

A    T    C    G

**6** Gene expression

RNA blot

Detection

© Infobase Publishing

Biotechnology. This technology consists of six basic steps: 1) digestion of DNA with restriction enzymes in order to isolate specific DNA fragments; 2) cloning of restriction fragments in circular bacterial minichromosomes to increase their numbers; 3) storing the fragments for further study in viral-based DNA libraries; 4) isolation and purification of DNA fragments from gene libraries using gel electrophoresis; 5) sequencing cloned DNA fragments; and 6) determining the expression profile of selected DNA clones using RNA blots and radioactive detection procedures.

## DNA Cloning

In 1973, scientists discovered that restriction enzymes (enzymes that can cut DNA at specific sites), DNA ligase (an enzyme that can join two pieces of DNA together), and bacterial plasmids could be used to clone DNA molecules. Plasmids are small (about 3,000 base pairs [bp]) circular minichromosomes that occur naturally in bacteria and are often exchanged between cells by passive diffusion. A bacterium is said to be transfected when it acquires a new plasmid. For bacteria, the main advantage to swapping plasmids is that they often carry antibiotic resistance genes, so that a cell sensitive to ampicillin can become resistant simply by acquiring the right plasmid. For scientists, plasmid swapping provided an ideal method for amplifying or cloning a specific piece of DNA.

The first cloning experiment used a plasmid from the bacterium *Escherichia coli* that was cut with the restriction enzyme *Eco*RI. The plasmid had a single *Eco*RI site, so the restriction enzyme simply opened the circular molecule. Foreign DNA, cut with the same restriction enzyme, was incubated with the plasmid. Because the plasmid and foreign DNA were both cut with *Eco*RI, the DNA could insert itself into the plasmid to form a hybrid, or recombinant plasmid, after which DNA ligase sealed the two together. The reaction mixture was added to a small volume of *E. coli* so that some of the cells could take up the recombinant plasmid before being transferred to a nutrient broth containing streptomycin. Only those cells carrying the recombinant plasmid, which contained an anti-streptomycin gene, could grow in the presence of this antibiotic. Each time the cells divided, the plasmid DNA was duplicated along with the main chromosome. After the cells had grown overnight, the foreign DNA had been amplified billions of times and was easily isolated for sequencing or expression studies. In this procedure, the plasmid is known as a cloning vector because it serves to transfer the foreign DNA into a cell.

## DNA Libraries

The basic cloning procedure described above not only provides a way to amplify a specific piece of DNA but can also be used to construct DNA libraries. In this case, however, the cloning vector is a bacteriophage called lambda. The lambda genome is double-stranded DNA of about 40,000 bp, much of which can be replaced by foreign DNA without sacrificing the ability of the virus to infect bacteria. This is the great advantage of lambda over a plasmid. Lambda can accommodate very long pieces of DNA, often long enough to contain an entire gene, whereas a plasmid cannot accommodate foreign DNA that is larger than 2,000 bp. Moreover, bacteriophage has the natural ability to infect bacteria, so that the efficiency of transfection is 100 times greater than it is for plasmids.

The construction of a DNA library begins with the isolation of genomic DNA and its digestion with a restriction enzyme to produce fragments of 1,000 to 10,000 bp. These fragments are ligated into lambda genomes, which are subjected to a packaging reaction to produce mature viral particles, most of which carry a different piece of the genomic DNA. This collection of viruses is called a genomic library and is used to study the structure and organization of specific genes. Clones from a library such as this contain the coding sequences, in addition to noncoding sequences such as introns, intervening sequences, promoters, and enhancers. An alternative form of a DNA library can be constructed by isolating messenger RNA (mRNA) from a specific cell type. This RNA is converted to the complementary DNA (cDNA) using an RNA-dependent DNA polymerase called reverse transcriptase. The cDNA is ligated to lambda genomes and packaged as for the genomic library. This collection of recombinant viruses is known as a cDNA library and contains genes that were being expressed by the cells when the mRNA was extracted. It does not include introns or controlling elements as these are lost during transcription and the processing that occurs in the cell to make mature mRNA. Thus a cDNA library is intended for the purpose of studying gene expression and the structure of the coding region only.

## Labeling Cloned DNA

Many of the procedures used in biotechnology were inspired by the events that occur during DNA replication (described above). This includes the labeling of cloned DNA for use as probes in expression studies, DNA sequencing, and PCR (described below). DNA replication involves duplicating one of the strands (the parent, or template strand) by linking nucleotides in an order specified by the template and depends on a large number of enzymes, the most important of which is DNA polymerase. This enzyme, guided by the template strand, constructs a daughter strand by linking nucleotides together. One such nucleotide is deoxyadenine triphosphate (dATP). Deoxyribonucleotides have a single hydroxyl group located at the 3′ carbon of the sugar group while the triphosphate is attached to the 5′ carbon.

The procedure for labeling DNA probes introduces radioactive nucleotides into a DNA molecule. This method supplies DNA polymerase with a single-stranded DNA template, a primer, and the four nucleotides in a buffered solution to induce in vitro replication. The daughter strand, which becomes the labeled probe, is made radioactive by including a $^{32}$P-labeled nucleotide in the reaction mix. The radioactive nucleotide is usually deoxy-cytosine triphosphate (dCTP) or dATP. The $^{32}$P is always part of the α (alpha) phosphate (the phosphate closest to the 5′ carbon), as this is the one used by the polymerase to form the phosphodiester bond between nucleotides. Nucleotides can also be labeled with a fluorescent dye molecule.

Single-stranded DNA hexamers (six bases long) are used as primers, and these are produced in such a way that they contain all possible permutations of four bases taken six at a time. Randomizing the base sequence for the primers ensures that there will be at least one primer site in a template that is only 50 bp long. Templates used in labeling reactions such as this are generally 100 to 800 bp long. This strategy of labeling DNA is known as random primer labeling.

## Gel Electrophoresis

This procedure is used to separate DNA and RNA fragments by size in a slab of agarose (highly refined agar) or polyacrylamide subjected to an electric field. Nucleic acids carry a negative charge and thus will migrate toward a positively charged electrode. The gel acts as a sieving medium that impedes the movement of the molecules. Thus, the rate at which the fragments migrate is a function of their size; small fragments migrate more rapidly than large fragments. The gel containing the samples is run submerged in a special pH-regulated solution, or buffer. Agarose gels are run horizontal, but DNA sequencing gels, made of polyacrylamide, are much bigger and are run in a vertical tank.

## DNA Sequencing

A sequencing reaction developed by the British biochemist Dr. Fred Sanger is a technique that takes its inspiration from the natural process of DNA replication. DNA polymerase requires a primer with a free 3′ hydroxyl group. The polymerase adds the first nucleotide to this group, and all subsequent bases are added to the 3′ hydroxyl of the previous base. Sequencing by the Sanger method is usually performed with the DNA cloned into a special sequencing plasmid. This simplifies the choice of the primers since their sequence can be derived from the known plasmid sequence. Once the primer binds to the primer site the cloned DNA may be replicated.

Sanger's innovation involved the synthesis of chain-terminating nucleotide analogues lacking the 3′ hydroxyl group. These analogues, also known as dideoxynucleotides (ddATP, ddCTP, ddGTP, and ddTTP), terminate the growth of the daughter strand at the point of insertion, and this can be used to determine the distance of each base on the daughter strand from the primer. These distances can be visualized by separating the Sanger reaction products on a polyacrylamide gel and then exposing the gel to X-ray film to produce an autoradiogram. The DNA sequence is read directly from this film, beginning with the smallest fragment at the bottom of the gel (the nucleotide closest to the primer) and ending with the largest fragment at the top. A hypothetical autoradiogram and the derived DNA sequence are shown in panel 5 of the figure on page 167. The smallest fragment in this example is the "C" nucleotide at the bottom of lane 3. The next nucleotide in the sequence is the "G" nucleotide in lane 4, then the "T" nucleotide in lane 2, and so on to the top of the gel.

Automated versions of the Sanger sequencing reaction use fluorescent-labeled dideoxynucleotides, each with a different color, so the sequence of the template can be recorded by a computer as the reaction mix passes a sensitive photocell. Machines such as this were used to sequence the human genome, a job that cost many millions of dollars and took years to complete. Recent advances in DNA-sequencing technology will make it possible to sequence the human genome in less than a week at a cost of $1,000.

## Gene Expression

The production of a genomic or cDNA library, followed by the sequencing of isolated clones, is a very powerful method for characterizing genes and the genomes from which they came. But the icing on the cake is the ability to determine the expression profile for a gene: That is, to determine which cells express the gene and exactly when the gene is turned on and off. Typical experiments may wish to determine the expression of specific genes in normal versus cancerous tissue or tissues obtained from groups of different ages. There are essentially three methods for doing this: RNA blotting, fluorescent in situ hybridization (FISH), and the polymerase chain reaction.

## RNA Blotting

This procedure consists of the following steps:

1. Extract mRNA from the cells or tissue of interest.
2. Fractionate (separate by size) the mRNA sample using gel electrophoresis.
3. Transfer the fractionated sample to a nylon membrane (the blotting step).
4. Incubate the membrane with a gene fragment (usually a cDNA clone) that has been labeled with a radioisotope.
5. Expose the membrane to X-ray film to visualize the signal.

The RNA is transferred from the gel to a nylon membrane using a vacuum apparatus or a simple dish containing a transfer buffer topped by a large stack of ordinary paper towels and a weight. The paper towels pull the transfer buffer through the gel, eluting the RNA from the gel and trapping it on the membrane. The location of specific mRNAs can be determined by hybridizing the membrane to a radiolabeled cDNA or genomic clone. The hybridization procedure involves placing the membrane in a buffer solution containing a labeled probe. During a long incubation period, the probe binds to the target sequence immobilized on the membrane. A-T and G-C base pairing (also known as hybridization) mediate the binding between the probe and target. The double-stranded molecule that is formed is a hybrid, being formed between the RNA target, on the membrane, and the DNA probe.

## Fluorescent In Situ Hybridization (FISH)

Studying gene expression does not always depend on RNA blots and membrane hybridization. DNA probes can be hybridized to DNA or RNA in situ, that is, while located within cells or tissue sections fixed on a microscope slide. In this case, the probe is labeled with a fluorescent dye molecule, rather than a radioactive isotope. The samples are then examined and photographed under a fluorescent microscope. FISH is an extremely powerful variation on RNA blotting. This procedure gives precise information regarding the identity of a cell that expresses a specific gene, information that usually cannot be obtained with membrane hybridization. Organs and tissues are generally composed of many different kinds of cells, which cannot be separated from one another using standard biochemical extraction procedures. Histological sections, however, show clearly the various cell types, and when subjected to FISH analysis, provide clear information as to which cells express specific genes. FISH is also used in clinical laboratories for the diagnosis of genetic abnormalities.

## Polymerase Chain Reaction (PCR)

PCR is simply repetitive DNA replication over a limited, primer-defined region of a suitable template. It provides a way of amplifying a short segment of DNA without going through the cloning procedures described above. The region defined by the primers is amplified to such an extent that it can be easily isolated for further study. The reaction exploits the fact that a DNA duplex, in a low-salt buffer, will melt (i.e., separate into two single strands) at 167°F (75°C) but will reanneal (rehybridize) at 98.6°F (37°C).

The reaction is initiated by melting the template, in the presence of primers and polymerase in a suitable buffer, cooling quickly to 98.6°F (37°C), and allowing sufficient time for the polymerase to replicate both strands of the template. The temperature is then increased to 167°F (75°C) to melt the newly formed duplexes and then cooled to 98.6°F (37°C). At the lower temperature, more primer will anneal to initiate another round of replication. The heating-cooling cycle is repeated 20 to 30 times, after which the reaction products are fractionated on an agarose gel, and the region containing the amplified fragment is cut out of the gel and purified for further study. The DNA polymerase used in these reactions is isolated from thermophilic bacteria that can withstand temperatures of 158°F (70°C) to 176°F (80°C). PCR applications are nearly limitless. It is used to amplify DNA from samples containing at times no more than a few cells. It is being used in the development of ultrafast DNA sequencers, identification of tissue samples in criminal investigations, amplification of ancient DNA obtained from fossils, and the identification of genes that are turned on or off during embryonic development or during cellular transformation (cancer formation).

## METHODS IN VIROLOGY

Scientists have developed many procedures over the years for studying viruses. The most basic methods provide ways to determine the amount of virus that is present in a sample, purification of the virus, and finally identification of the isolated specimen.

## Determining the Number of Viral Particles

PCR can be used to determine the number of viral genomes that are present in a sample. The accuracy of this procedure depends on the purity of the sample and whether there is more than one viral species present or not. Influenza viruses can be quantified with a hemagglutination assay. The hemagglutinin envelope spike of the flu viruses has the ability to agglutinate human red blood cells (i.e., the RBCs form clusters around the viral particles), and this phenomenon can be calibrated to estimate the number of viral particles present in a sample. Viral particles can also be counted directly under an electron microscope by comparison with a standard suspension of latex beads of a similar size. The most common biological method for quantifying viruses is called the plaque assay. This assay is based on the fact that viruses either lyse cells directly or that infected cells eventually die and the membrane ruptures spontaneously. Host cells are spread on a culture plate and allowed to cover the entire surface. The growth of the cells causes the plate to become slightly opaque. Next, a thin coat of melted agar, containing a suspension of the viruses, is spread over the top of the plate. After a few days, the virus will have infected and killed many of the host cells. The areas where the cells have died and broken open become transparent, and these areas appear as small circular clear areas (the plaques) over the surface of the dish. Counting these plaques and then multiplying by the dilution factor provides a reasonably accurate estimate of the number of viral particles in the sample.

## Purification of Viruses

Pure virus preparations are required for basic research and as starting material for the production of a vaccine. Viruses are usually purified from large volumes of tissue culture medium, body fluids, or infected cells. The initial step involves concentration of the viral particles with ethanol or by spinning the preparation in an ultracentrifuge. The conditions within the centrifuge tube can be adjusted so that anything larger than a virus will remain in solution while the viruses form a pellet at the bottom of the tube. Ultracentrifugation may be repeated several times, with slightly different solutions, until the virus is nearly free of cellular material. Naked icosahedral viruses are easier to purify than any virus that has an envelope.

## Identification of Viruses

Initial identification is carried out under an electron microscope. This will provide the basic information regarding the type of capsid, its size, and whether it has an envelope or not. The next step is to determine the type of genome, as to whether it is DNA or RNA. This can be done with a spectrophotometer; since these molecules absorb light at different frequencies it is relatively easy to distinguish between the two. It can also be done enzymatically, by digesting a sample with RNase or DNase. Additional studies can be conducted by isolating the proteins and identifying them using a procedure called Western blot analysis. This procedure is similar to the RNA blots described previously. Viral proteins are separated on an electrophoresis gel, transferred to a membrane, and probed with labeled antibodies to known viral proteins.

## UNDERSTANDING CLINICAL TRIALS

Clinical trials are conducted in four phases and are always preceded by research conducted on experimental animals such as mice, rats, or monkeys. The format for preclinical research is informal; it is conducted in a variety of research labs around the world, with the results being published in scientific journals. Formal approval from a governmental regulatory body is not required.

## Phase I Clinical Trial

Pending the outcome of the preclinical research, investigators may apply for permission to try the experiments on human subjects. Applications in the United States are made to the Food and Drug Administration (FDA), the National Institutes of Health (NIH), and the Recombinant DNA Advisory Committee (RAC). RAC was set up by NIH to monitor any research, including clinical trials, dealing with cloning, recombinant DNA, or gene therapy. Phase I trials are conducted on a small number of adult volunteers, usually between two and 20, who have given informed consent. That is, the investigators explain the procedure, the possible outcomes, and especially the dangers associated with the procedure before the subjects sign a consent form. The purpose of the Phase I trial is to determine the overall effect the treatment has on humans. A treatment that works well in monkeys or mice may not work at all on humans. Similarly, a treatment that appears safe in lab animals may be toxic, even deadly, when given to humans. Since most clinical trials are testing a new drug of some kind, the first priority is to determine a safe dosage for humans. Consequently, subjects in the Phase I trial are given a range of doses, all of which, even the high dose, are less than the highest dose given to experimental animals.

If the results from the Phase I trial are promising, the investigators may apply for permission to proceed to Phase II.

## Phase II Clinical Trial

Having established the general protocol, or procedure, the investigators now try to replicate the encouraging results from Phase I, but with a much larger number of subjects (100 to 300). Only with a large number of subjects is it possible to prove the treatment has an effect. In addition, dangerous side effects may have been missed in Phase I because of a small sample size. The results from Phase II will determine how safe the procedure is and whether it works or not. If the statistics show the treatment is effective and toxicity is low, the investigators may apply for permission to proceed to Phase III.

## Phase III Clinical Trial

Based on Phase II results, the procedure may look very promising, but before it can be used as a routine treatment it must be tested on thousands of patients at a variety of research centers. This is the expensive part of bringing a new drug or therapy to market, costing millions, sometimes billions, of dollars. It is for this reason that Phase III clinical trials invariably have the financial backing of large pharmaceutical or biotechnology companies. If the results of the Phase II trial are confirmed in Phase III, the FDA will approve the use of the drug for routine treatment. The use of the drug or treatment now passes into an informal Phase IV trial.

## Phase IV Clinical Trial

Even though the treatment has gained formal approval, its performance is monitored for very long-term effects, sometimes stretching on for 10 to 20 years. In this way, the FDA retains the power to recall the drug long after it has become a part of standard medical procedure. It can happen that in the long term, the drug costs more than an alternative, in which case, health insurance providers may refuse to cover the cost of the treatment.

## GENE AND PROTEIN NOMENCLATURE

Scientists who were, in effect, probing around in the dark have discovered many genes and their encoded proteins. Once discovered, the new genes or proteins had to be named. Usually the "name" is nothing more than a lab-book code or an acronym suggested by the system under study at the time. Sometimes it turns out, after further study, that the function observed in the original study is a minor aspect of the gene's role in the cell. It is for this reason that gene and protein names sometimes seem absurd and poorly chosen.

In 2003, an International Committee on Standardized Genetic Nomenclature agreed to unify the rules and guidelines for gene and protein names for the mouse and rat. Similar committees have attempted to standardize gene-naming conventions for human, frog, zebrafish, and yeast genes. In general, the gene name is expected to be brief and to begin with a lowercase letter unless it is a person's name. The gene symbols are acronyms taken from the gene name and are expected to be three to five characters long and not more than 10. The symbols must be written with Roman letters and Arabic numbers. The same symbol is used for orthologs (i.e., the same gene) among different species, such as human, mouse, or rat. Thus the gene sonic hedgehog is symbolized as shh, and the gene myelocytomatosis is symbolized as myc.

Unfortunately, the various committees were unable to agree on a common presentation for the gene and protein symbols. A human gene symbol, for example, is italicized, uppercase letters, and the protein is uppercase and not italicized. A frog gene symbol is lowercase and the protein is uppercase, while neither is italicized. Thus the myc gene and its protein, for example, are written as *MYC* and MYC in humans, myc and MYC in frogs, and *Myc* and Myc in mice and rats. The latter convention, *Myc* and Myc, is used throughout the New Biology set, regardless of the species.

## WEIGHTS AND MEASURES

The following table presents some common weights, measures, and conversions that appear in this book and other volumes of the New Biology set.

| QUANTITY | EQUIVALENT |
|----------|-----------|
| Length | 1 meter (m) = 100 centimeters (cm) = 1.094 yards = 39.37 inches<br>1 kilometer (km) = 1,000 m = 0.62 miles<br>1 foot = 30.48 cm<br>1 inch = 1/12 foot = 2.54 cm<br>1 cm = 0.394 inch = $10^{-2}$ (or 0.01) m<br>1 millimeter (mm) = $10^{-3}$ m<br>1 micrometer (μm) = $10^{-6}$ m<br>1 nanometer (nm) = $10^{-9}$ m<br>1 Ångström (Å) = $10^{-10}$ m |
| Mass | 1 gram (g) = 0.0035 ounce<br>1 pound = 16 ounces = 453.6 grams<br>1 kilogram (kg) = 2.2 pounds (lb)<br>1 milligram (mg) = $10^{-3}$ g<br>1 microgram (μg) = $10^{-6}$ g |
| Volume | 1 liter (l) = 1.06 quarts (US) = 0.264 gallon (US)<br>1 quart (US) = 32 fluid ounces = 0.95 liter<br>1 milliliter (ml) = $10^{-3}$ liter = 1 cubic centimeter (cc) |
| Temperature | °C = 5/9 (°F - 32)<br>°F = (9/5 × °C) + 32 |
| Energy | Calorie = the amount of heat needed to raise the temperature of 1 gram of water by 1°C.<br>Kilocalorie = 1,000 calories. Used to describe the energy content of foods. |

 **Glossary**

**acetyl** A chemical group derived from acetic acid that is important in energy metabolism and for the modification of proteins.

**acetylcholine** A neurotransmitter released at axonal terminals by cholinergic neurons, found in the central and peripheral nervous systems and released at the vertebrate neuromuscular junction.

**acetyl-CoA** A water-soluble molecule, coenzyme A (CoA) that carries acetyl groups in cells.

**acid** A substance that releases protons when dissolved in water; carries a net negative charge.

**actin filament** A protein filament formed by the polymerization of globular actin molecules; forms the cytoskeleton of all eukaryotes and part of the contractile apparatus of skeletal muscle.

**action potential** A self-propagating electrical impulse that occurs in the membranes of neurons, muscles, photoreceptors, and hair cells of the inner ear.

**active transport** Movement of molecules across the cell membrane, using the energy stored in ATP.

**adenylate cyclase** A membrane-bound enzyme that catalyzes the conversion of ATP to cyclic AMP; an important component of cell-signaling pathways.

**adherens junction** A cell junction in which the cytoplasmic face of the membrane is attached to actin filaments.

**adipocyte** A fat cell.

**adrenaline (epinephrine)** A hormone released by chromaffin cells in the adrenal gland; prepares an animal for extreme activity by increasing the heart rate and blood sugar levels.

**adult stem cells**   Stem cells isolated from adult tissues, such as bone marrow or epithelium.

**aerobic**   Refers to a process that either requires oxygen or occurs in its presence.

**agar**   A polysaccharide isolated from seaweed that forms a gel when boiled in water and cooled to room temperature; used by microbiologists as a solid culture medium for the isolation and growth of bacteria and fungi.

**agarose**   A purified form of agar that is used to fractionate (separate by size) biomolecules.

**allele**   An alternate form of a gene. Diploid organisms have two alleles for each gene, located at the same locus (position) on homologous chromosomes.

**allogeneic transplant**   A cell, tissue, or organ transplant from an unrelated individual.

**alpha helix**   A common folding pattern of proteins in which a linear sequence of amino acids twists into a right-handed helix stabilized by hydrogen bonds.

**amino acid**   An organic molecule containing amino and carboxyl groups that is a building block of protein.

**aminoacyl tRNA**   An amino acid linked by its carboxyl group to a hydroxyl group on tRNA.

**aminoacyl-tRNA synthetase**   An enzyme that attaches the correct amino acid to a tRNA.

**amino terminus**   The end of a protein or polypeptide chain that carries a free amino group.

**amphipathic**   Having both hydrophilic and hydrophobic regions, as in a phospholipid.

**anabolism**   A collection of metabolic reactions in a cell whereby large molecules are made from smaller ones.

**anaerobic**   A cellular metabolism that does not depend on molecular oxygen.

**anaphase**   A mitotic stage in which the two sets of chromosomes move away from each other toward opposite spindle poles.

**anchoring junction**   A cell junction that attaches cells to each other.

**angiogenesis**   Sprouting of new blood vessels from preexisting ones.

**angstrom**   A unit of length, equal to $10^{-10}$ meter or 0.1 nanometer (nM), that is used to measure molecules and atoms.

**anterior**   A position close to or at the head end of the body.

**antibiotic**   A substance made by bacteria, fungi, and plants that is toxic to microorganisms. Common examples are penicillin and streptomycin.

**antibody**   A protein made by B cells of the immune system in response to invading microbes.

**anticodon**   A sequence of three nucleotides in tRNA that is complementary to a messenger RNA codon.

**antigen**   A molecule that stimulates an immune response, leading to the formation of antibodies.

**antigen-presenting cell**   A cell of the immune system, such as a monocyte, that presents pieces of an invading microbe (the antigen) to lymphocytes.

**antiparallel**   The relative orientation of the two strands in a DNA double helix; the polarity of one strand is oriented in the opposite direction to the other.

**antiporter**   A membrane carrier protein that transports two different molecules across a membrane in opposite directions.

**apoptosis**   Regulated or programmed form of cell death that may be activated by the cell itself or by the immune system to force cells to commit suicide when they become infected with a virus or bacterium.

**archaea**   The archaea are prokaryotes that are physically similar to bacteria (both lack a nucleus and internal organelles), but they have retained a primitive biochemistry and physiology that would have been commonplace 2 billion years ago.

**asexual reproduction**   The process of forming new individuals without gametes or the fertilization of an egg by a sperm. Individuals produced this way are identical to the parent and referred to as a clone.

**aster**   The star-shaped arrangement of microtubules that is characteristic of a mitotic or meiotic spindle.

**ATP (adenosine triphosphate)**   A nucleoside consisting of adenine, ribose, and three phosphate groups that is the main carrier of chemical energy in the cell.

**ATPase**   Any enzyme that catalyzes a biochemical reaction by extracting the necessary energy from ATP.

**ATP synthase**   A protein located in the inner membrane of the mitochondrion that catalyzes the formation of ATP from ADP and inorganic phosphate using the energy supplied by the electron transport chain.

**autogeneic transplant**   A patient receives a transplant of his or her own tissue.

**autologous**   Refers to tissues or cells derived from the patient's own body.

**autoradiograph (autoradiogram)**   X-ray film that has been exposed to X-rays or to a source of radioactivity; used to visualize internal structures of the body and radioactive signals from sequencing gels and DNA or RNA blots.

**autosome**   Any chromosome other than a sex chromosome.

**axon**   A long extension of a neuron's cell body that transmits an electrical signal to other neurons.

**axonal transport**   The transport of organelles, such as Golgi vesicles, along an axon to the axonal terminus. Transport also flows from the terminus to the cell body.

**bacteria**   One of the most ancient forms of cellular life (the other is the archaea). Bacteria are prokaryotes, and some are known to cause disease.

**bacterial artificial chromosome (BAC)**   A cloning vector that accommodates DNA inserts of up to 1 million base pairs.

**bacteriophage**   A virus that infects bacteria. Bacteriophages were used to prove that DNA is the cell's genetic material and are now used as cloning vectors.

**base**   A substance that can accept a proton in solution. The purines and pyrimidines in DNA and RNA are organic bases and are often referred to simply as bases.

**base pair**   Two nucleotides in RNA or DNA that are held together by hydrogen bonds. Adenine bound to thymine or guanine bound to cytosine are examples of base pairs

**B cell (B lymphocyte)**   A white blood cell that makes antibodies and is part of the adaptive immune response.

**benign**   Tumors that grow to a limited size and do not spread to other parts of the body.

**beta sheet**  Common structural motif in proteins in which different strands of the protein run alongside one another and are held together by hydrogen bonds.

**biopsy**  The removal of cells or tissues for examination under a microscope. When only a sample of tissue is removed, the procedure is called an incisional biopsy or core biopsy. When an entire lump or suspicious area is removed, the procedure is called an excisional biopsy. When a sample of tissue or fluid is removed with a needle, the procedure is called a needle biopsy or fine-needle aspiration.

**biosphere**  The world of living organisms

**biotechnology**  A set of procedures that are used to study and manipulate genes and their products.

**blastomere**  A cell formed by the cleavage of a fertilized egg. Blastomeres are the totipotent cells of the early embryo.

**blotting**  A technique for transferring DNA (southern blotting), RNA (northern blotting), or proteins (western blotting) from an agarose or polyacrylamide gel to a nylon membrane.

**BRCA1 (breast cancer gene 1)**  A gene on chromosome 17 that may be involved in regulating the cell cycle. A person who inherits an altered version of the BRCA1 gene has a higher risk of getting breast, ovarian, or prostate cancer.

**BRCA2 (breast cancer gene 2)**  A gene on chromosome 13 that, when mutated, increases the risk of getting breast, ovarian, or prostate cancer.

**budding yeast**  The common name for the baker's yeast *Saccharomyces cerevisiae,* a popular experimental organism that reproduces by budding off a parental cell.

**buffer**  A pH-regulated solution with a known electrolyte (salt) content; used in the isolation, manipulation, and storage of biomolecules and medicinal products.

**cadherin**  Belongs to a family of proteins that mediates cell-cell adhesion in animal tissues.

**calorie**  A unit of heat. One calorie is the amount of heat needed to raise the temperature of one gram of water by 1°C. kilocalories (1,000 calories) are used to describe the energy content of foods.

**capsid**   The protein coat of a virus, formed by autoassembly of one or more proteins into a geometrically symmetrical structure.

**carbohydrate**   A general class of compounds that includes sugars, containing carbon, hydrogen, and oxygen.

**carboxyl group**   A carbon atom attached to an oxygen and a hydroxyl group

**carboxyl terminus**   The end of a protein containing a carboxyl group.

**carcinogen**   A compound or form of radiation that can cause cancer.

**carcinogenesis**   The formation of a cancer.

**carcinoma**   Cancer of the epithelium, representing the majority of human cancers.

**cardiac muscle**   Muscle of the heart; composed of myocytes that are linked together in a communication network based on free passage of small molecules through gap junctions.

**caspase**   A protease involved in the initiation of apoptosis.

**catabolism**   Enzyme regulated breakdown of large molecules for the extraction of chemical-bond energy. Intermediate products are called catabolites.

**catalyst**   A substance that lowers the activation energy of a reaction.

**CD28**   Cell-surface protein located in T-cell membranes, necessary for the activation of T-cells by foreign antigens.

**cDNA (complementary DNA)**   DNA that is synthesized from mRNA, thus containing the complementary sequence; cDNA contains coding sequence, but not the regulatory sequences that are present in the genome. Labeled probes are made from cDNA for the study of gene expression.

**cell adhesion molecule (CAM)**   A cell surface protein that is used to connect cells to one another.

**cell body**   The main part of a cell containing the nucleus, Golgi complex, and endoplasmic reticulum; used in reference to neurons that have long processes (dendrites and axons) extending some distance from the nucleus and cytoplasmic machinery.

**cell coat**   (see **glycocalyx**)

**cell-cycle control system**   A team of regulatory proteins that governs progression through the cell cycle.

**cell-division-cycle gene (*cdc* gene)**   A gene that controls a specific step in the cell cycle.

**cell fate**   The final differentiated state that a pluripotent embryonic cell is expected to attain.

**cell-medicated immune response**   Activation of specific cells to launch an immune response against an invading microbe.

**cell nuclear transfer**   Animal cloning technique whereby a somatic cell nucleus is transferred to an enucleated oocyte; synonymous with somatic cell nuclear transfer.

**celsius**   A measure of temperature. This scale is defined such that 0°C is the temperature at which water freezes and 100°C is the temperature at which water boils.

**central nervous system (CNS)**   That part of a nervous system that analyzes signals from the body and the environment. In animals, the CNS includes the brain and spinal cord.

**centriole**   A cylindrical array of microtubules that is found at the center of a centrosome in animal cells.

**centromere**   A region of a mitotic chromosome that holds sister chromatids together. Microtubules of the spindle fiber connect to an area of the centromere called the kinetochore.

**centrosome**   Organizes the mitotic spindle and the spindle poles; in most animal cells it contains a pair of centrioles.

**chiasma (plural chiasmata)**   An X-shaped connection between homologous chromosomes that occurs during meiosis I, representing a site of crossing-over, or genetic exchange between the two chromosomes.

**chromatid**   A duplicate chromosome that is still connected to the original at the centromere. The identical pair are called sister chromatids.

**chromatin**   A complex of DNA and proteins (histones and non-histones) that forms each chromosome and is found in the nucleus of all eukaryotes. Decondensed and threadlike during interphase.

**chromatin condensation**   Compaction of different regions of interphase chromosomes that is mediated by the histones.

**chromosome**   One long molecule of DNA that contains the organism's genes. In prokaryotes, the chromosome is circular and naked; in eukaryotes, it is linear and complexed with histone and nonhistone proteins.

**chromosome condensation**   Compaction of entire chromosomes in preparation for cell division.

**clinical breast exam**   An exam of the breast performed by a physician to check for lumps or other changes.

**cnidoblast**   A stinging cell found in the Cnidarians (jellyfish).

**cyclic adenosine monophosphate (cAMP)**   A second messenger in a cell-signaling pathway that is produced from ATP by the enzyme adenylate cyclase.

**cyclin**   A protein that activates protein kinases (cyclin-dependent protein kinases, or Cdk) that control progression from one stage of the cell cycle to another.

**cytochemistry**   The study of the intracellular distribution of chemicals.

**cytochrome**   Colored, iron-containing protein that is part of the electron transport chain.

**cytotoxic T cell**   A T lymphocyte that kills infected body cells.

**dendrite**   An extension of a nerve cell that receives signals from other neurons.

**dexrazoxane**   A drug used to protect the heart from the toxic effects of anthracycline drugs such as doxorubicin. It belongs to the family of drugs called chemoprotective agents.

**dideoxynucleotide**   A nucleotide lacking the 2' and 3' hydroxyl groups.

**dideoxy sequencing**   A method for sequencing DNA that employs dideoxyribose nucleotides; also known as the Sanger sequencing method, after Fred Sanger, a chemist who invented the procedure in 1976.

**diploid**   A genetic term meaning two sets of homologous chromosomes, one set from the mother and the other from the father. Thus, diploid organisms have two versions (alleles) of each gene in the genome.

**DNA (deoxyribonucleic acid)**   A long polymer formed by linking four different kinds of nucleotides together likes beads on a string. The sequence of nucleotides is used to encode an organism's genes.

**DNA helicase**   An enzyme that separates and unwinds the two DNA strands in preparation for replication or transcription.

**DNA library**   A collection of DNA fragments that are cloned into plasmids or viral genomes.

**DNA ligase**   An enzyme that joins two DNA strands together to make a continuous DNA molecule.

**DNA microarray**   A technique for studying the simultaneous expression of a very large number of genes.

**DNA polymerase**   An enzyme that synthesizes DNA using one strand as a template.

**DNA primase**   An enzyme that synthesizes a short strand of RNA that serves as a primer for DNA replication.

**dorsal**   The backside of an animal; also refers to the upper surface of anatomical structures, such as arms or wings.

**dorsalventral**   The body axis running from the backside to the frontside or the upperside to the underside of a structure.

**double helix**   The three-dimensional structure of DNA in which the two strands twist around each other to form a spiral.

**doxorubicin**   An anticancer drug that belongs to a family of antitumor antibiotics.

***Drosophila melanogaster***   Small species of fly, commonly called a fruit fly, that is used as an experimental organism in genetics, embryology, and gerontology.

**ductal carcinoma in situ (DCIS)**   Abnormal cells that involve only the lining of a breast duct. The cells have not spread outside the duct to other tissues in the breast; also called intraductal carcinoma.

**dynein**   A motor protein that is involved in chromosome movements during cell division.

**dysplasia**   Disordered growth of cells in a tissue or organ, often leading to the development of cancer.

**ectoderm**   An embryonic tissue that is the precursor of the epidermis and the nervous system.

**electrochemical gradient**   A differential concentration of an ion or molecule across the cell membrane that serves as a source of potential energy and may polarize the cell electrically.

**electron microscope**   A microscope that uses electrons to produce a high-resolution image of the cell.

**electrophoresis**   The movement of a molecule, such as protein, DNA, or RNA, through an electric field. In practice, the molecules migrate through a slab of agarose or polyacrylamide that is immersed in a special solution and subjected to an electric field.

**elution**   To remove one substance from another by washing it out with a buffer or solvent.

**embryogenesis**   The development of an embryo from a fertilized egg.

**embryonic stem cell (ES cell)**   A pluripotent cell derived from the inner cell mass (the cells that give rise to the embryo instead of the placenta) of a mammalian embryo.

**endocrine cell**   A cell that is specialized for the production and release of hormones. Such cells make up hormone-producing tissue such as the pituitary gland or gonads.

**endocytosis**   Cellular uptake of material from the environment by invagination of the cell membrane to form a vesicle called an endosome. The endosome's contents are made available to the cell after it fuses with a lysosome.

**endoderm**   An embryonic tissue layer that gives rise to the gut.

**endoplasmic reticulum (ER)**   Membrane-bounded chambers that are used to modify newly synthesized proteins with the addition of sugar molecules (glycosylation). When finished, the glycosylated proteins are sent to the Golgi apparatus in exocytotic vesicles.

**enhancer**   A DNA-regulatory sequence that provides a binding site for transcription factors capable of increasing the rate of transcription for a specific gene; often located thousands of base pairs away from the gene it regulates.

**enveloped virus**   A virus containing a capsid that is surrounded by a lipid bilayer originally obtained from the membrane of a previously infected cell.

**enzyme**   A protein or RNA that catalyzes a specific chemical reaction.

**epidermis**   The epithelial layer, or skin, that covers the outer surface of the body.

**ER marker sequence**   The amino terminal sequence that directs proteins to enter the endoplasmic reticulum (ER). This sequence is removed once the protein enters the ER.

**erythrocyte**   A red blood cell that contains the oxygen-carrying pigment hemoglobin; used to deliver oxygen to cells in the body.

***Escherichia coli* (E. coli)**   Rod-shape, gram-negative bacterium that inhabits the intestinal tract of most animals and is used as an experimental organism by geneticists and biomedical researchers.

**euchromatin**   Lightly staining portion of interphase chromatin, in contrast to the darkly staining heterochromatin (condensed chromatin). Euchromatin contains most, if not all, of the active genes.

**eukaryote (eucaryote)** A cell containing a nucleus and many membrane-bounded organelles. All life-forms, except bacteria and viruses, are composed of eukaryote cells.

**exocytosis** The process by which molecules are secreted from a cell. Molecules to be secreted are located in Golgi-derived vesicles that fuse with the inner surface of the cell membrane, depositing the contents into the intercellular space.

**exon** Coding region of a eukaryote gene that is represented in messenger RNA and thus directs the synthesis of a specific protein.

**expression studies** Examination of the type and quantity of mRNA or protein that is produced by cells, tissues, or organs.

**fat** A lipid material, consisting of triglycerides (fatty acids bound to glycerol), that is stored adipocytes as an energy reserve.

**fatty acid** A compound that has a carboxylic acid attached to a long hydrocarbon chain. A major source of cellular energy and a component of phospholipids.

**fertilization** The fusion of haploid male and female gametes to form a diploid zygote.

**fibroblast** The cell type that, by secreting an extracellular matrix, gives rise to the connective tissue of the body.

**Filopodium** A fingerlike projection of a cell's cytoplasmic membrane, commonly observed in amoeba and embryonic nerve cells.

**filter hybridization** The detection of specific DNA or RNA molecules, fixed on a nylon filter (or membrane), by incubating the filter with a labeled probe that hybridizes to the target sequence; also known as membrane hybridization.

**fixative** A chemical that is used to preserve cells and tissues. Common examples are formaldehyde, methanol, and acetic acid.

**flagellum (plural flagella)** Whiplike structure found in prokaryotes and eukaryotes that are used to propel cells through water.

**fluorescein** Fluorescent dye that produces a green light when illuminated with ultraviolet or blue light.

**fluorescent dye** A dye that absorbs UV or blue light and emits light of a longer wavelength, usually as green or red light.

**fluorescent in situ hybridization (FISH)** A procedure for detecting the expression of a specific gene in tissue sections or smears through the use of DNA probes labeled with a fluorescent dye.

**fluorescent microscope**    A microscope that is equipped with special filters and a beam splitter for the examination of tissues and cells stained with a fluorescent dye.

**follicle cell**    Cells that surround and help feed a developing oocyte.

**$G_0$**    G "zero" refers to a phase of the cell cycle; state of withdrawal from the cycle as the cell enters a resting or quiescent stage; occurs in differentiated body cells, as well as in developing oocytes.

**$G_1$**    Gap 1 refers to the phase of the cell cycle that occurs just after mitosis and before the next round of DNA synthesis.

**$G_2$**    The Gap 2 phase of the cell cycle follows DNA replication and precedes mitosis.

**gap junction**    A communication channel in the membranes of adjacent cells that allows free passage of ions and small molecules.

**gel electrophoresis**    A procedure that is used to separate biomolecules by forcing them to migrate through a gel matrix (agarose or polyacrylamide) subjected to an electric field.

**gene**    A region of the DNA that specifies a specific protein or RNA molecule that is handed down from one generation to the next. This region includes both the coding, noncoding, and regulatory sequences.

**gene regulatory protein**    Any protein that binds to DNA and thereby affects the expression of a specific gene.

**gene repressor protein**    A protein that binds to DNA and blocks transcription of a specific gene.

**gene therapy**    A method for treating disease whereby a defective gene, causing the disease, is either repaired, replaced, or supplemented with a functional copy.

**genetic code**    A set of rules that assigns a specific DNA or RNA triplet, consisting of a three-base sequence, to a specific amino acid.

**genome**    All of the genes that belong to a cell or an organism.

**genomic library**    A collection of DNA fragments, obtained by digesting genomic DNA with a restriction enzyme, that are cloned into plasmid or viral vectors.

**genomics**    The study of DNA sequences and their role in the function and structure of an organism.

**genotype**    The genetic composition of a cell or organism.

**germ cell**    Cells that develop into gametes, either sperm or oocytes.

**glucose**    Six-carbon monosaccharide (sugar) that is the principal source of energy for many cells and organisms; stored as glycogen in animal cells and as starch in plants. Wood is an elaborate polymer of glucose and other sugars.

**glycerol**    A three-carbon alcohol that is an important component of phospholipids.

**glycocalyx**    A molecular "forest," consisting of glycosylated proteins and lipids, that covers the surface of every cell. The glycoproteins and glycolipids, carried to the cell membrane by Golgi-derived vesicles, have many functions including the formation of ion channels, cell-signaling receptors, and transporters.

**glycogen**    A polymer of glucose, used to store energy in an animal cell.

**glycolysis**    The degradation of glucose with production of ATP.

**glycoprotein**    Any protein that has a chain of glucose molecules (oligosaccharide) attached to some of the amino acid residues.

**glycosylation**    The process of adding one or more sugar molecules to proteins or lipids.

**glycosyltransferase**    An enzyme in the Golgi complex that adds glucose to proteins.

**Golgi complex (Golgi apparatus)**    Membrane-bounded organelle in eukaryote cells that receives glycoproteins from the ER, which are modified and sorted before being sent to their final destination. The Golgi complex is also the source of glycolipids that are destined for the cell membrane. The glycoproteins and glycolipids leave the Golgi by exocytosis. This organelle is named after the Italian histologist Camillo Golgi, who discovered it in 1898.

**Gram stain**    A bacterial stain that detects different species of bacteria based on the composition of their cell wall. Bacteria that retain the Gram stain are colored blue (Gram positive), whereas those that do not are colored orange (Gram negative).

**granulocyte**    A type of white blood cell that includes the neutrophils, basophils, and eosinophils.

**growth factor**    A small protein (polypeptide) that can stimulate cells to grow and proliferate.

**haploid**    Having only one set of chromosomes; a condition that is typical in gametes, such as sperm and eggs.

**HeLa cell** A tumor-derived cell line, originally isolated from a cancer patient in 1951; currently used by many laboratories to study the cell biology of cancer and carcinogenesis.

**helix-loop-helix** A structural motif common to a group of gene-regulatory proteins.

**helper T cell** A type of T lymphocyte that helps stimulate B cells to make antibodies directed against a specific microbe or antigen.

**hemoglobin** An iron-containing protein complex, located in red blood cells, that picks up oxygen in the lungs and carries it to other tissues and cells of the body.

**hemopoiesis** Production of blood cells, occurring primarily in the bone marrow.

**hematopoietic** Refers to cells, derived form the bone marrow, that give rise to red and white blood cells.

**hematopoietic stem cell transplantation (HSCT)** The use of stem cells isolated from the bone marrow to treat leukemia and lymphoma.

**hepatocyte** A liver cell.

**heterochromatin** A region of a chromosome that is highly condensed and transcriptionally inactive.

**histochemistry** The study of chemical differentiation of tissues.

**histology** The study of tissues.

**histone** Small nuclear proteins, rich in the amino acids arginine and lysine, that form the nucleosome in eukaryote nuclei, a beadlike structure that is a major component of chromatin.

**HIV** The human immunodeficiency virus that is responsible for AIDS.

**homolog** One of two or more genes that have a similar sequence and are descended from a common ancestor gene.

**homologous** Organs or molecules that are similar in structure because they have descended from a common ancestor; used primarily in reference to DNA and protein sequences.

**homologous chromosomes** Two copies of the same chromosome, one inherited from the mother and the other from the father.

**hormone** A signaling molecule, produced and secreted by endocrine glands; usually released into general circulation for coordination of an animal's physiology.

**housekeeping gene**    A gene that codes for a protein that is needed by all cells, regardless of the cell's specialization. Genes encoding enzymes involved in glycolysis and Krebs cycle are common examples.

**hybridization**    A term used in molecular biology (recombinant DNA technology) meaning the formation a double-stranded nucleic acid through complementary base-pairing; a property that is exploited in filter hybridization; a procedure that is used to screen gene libraries and to study gene structure and expression.

**hydrolysis**    The breaking of a covalent chemical bond with the subsequent addition of a molecule of water.

**hydrophilic**    A polar compound that mixes readily with water.

**hydrophobic**    A nonpolar molecule that dissolves in fat and lipid solutions, but not in water.

**hydroxyl group (-OH)**    Chemical group consisting of oxygen and hydrogen that is a prominent part of alcohol.

**image analysis**    A computerized method for extracting information from digitized microscopic images of cells or cell organelles.

**immunofluorescence**    Detection of a specific cellular protein with the aid of a fluorescent dye that is coupled to an antibody.

**immunoglobulin (Ig)**    An antibody made by B cells as part of the adaptive immune response.

**incontinence**    Inability to control the flow of urine from the bladder (urinary incontinence) or the escape of stool from the rectum (fecal incontinence).

**insertional mutagenesis**    Damage suffered by a gene when a virus or a jumping gene inserts itself into a chromosome.

**in situ hybridization**    A method for studying gene expression, whereby a labeled cDNA or RNA probe hybridizes to a specific mRNA in intact cells or tissues. The procedure is usually carried out on tissue sections or smears of individual cells.

**insulin**    Polypeptide hormone secreted by $\beta$ (beta) cells in the vertebrate pancreas. Production of this hormone is regulated directly by the amount of glucose that is in the blood.

**interleukin**    A small protein hormone, secreted by lymphocytes, to activate and coordinate the adaptive immune response.

**interphase**    The period between each cell division, which includes the $G_1$, S, and $G_2$ phases of the cell cycle.

**intron**    A section of a eukaryotic gene that is noncoding. It is transcribed but does not appear in the mature mRNA.

**in vitro**    Refers to cells growing in culture or a biochemical reaction occurring in a test tube (Latin for "in glass").

**in vivo**    A biochemical reaction, or a process, occurring in living cells or a living organism (Latin for "in life").

**ion**    An atom that has gained or lost electrons, thus acquiring a charge. Common examples are $Na^+$ and $Ca^{++}$ ions.

**ion channel**    A transmembrane channel that allows ions to diffuse across the membrane down their electrochemical gradient.

**ischemia**    An inadequate supply of blood to a part of the body caused by degenerative vascular disease.

**Jak-STAT signaling pathway**    One of several cell signaling pathways that activates gene expression. The pathway is activated through cell surface receptors and cytoplasmic Janus kinases (Jaks) and signal transducers and activators of transcription (STATs).

**karyotype**    A pictorial catalogue of a cell's chromosomes, showing their number, size, shape, and overall banding pattern.

**keratin**    Proteins produced by specialized epithelial cells called keratinocytes. Keratin is found in hair, fingernails, and feathers.

**kilometer**    1,000 meters, which is equal to 0.621 miles.

**kinesin**    A motor protein that uses energy obtained from the hydrolysis of ATP to move along a microtubule.

**kinetochore**    A complex of proteins that forms around the centromere of mitotic or meiotic chromosomes, providing an attachment site for microtubules. The other end of each microtubule is attached to a chromosome.

**Krebs cycle (citric acid cycle)**    The central metabolic pathway in all eukaryotes and aerobic prokaryotes; discovered by the German chemist Hans Krebs in 1937. The cycle oxidizes acetyl groups derived from food molecules. The end products are $CO_2$, $H_2O$, and high-energy electrons, which pass via NADH and FADH2 to the respiratory chain. In eukaryotes, the Krebs cycle is located in the mitochondria.

**labeling reaction**    The addition of a radioactive atom or fluorescent dye to DNA or RNA for use as a probe in filter hybridization.

**lagging strand**    One of the two newly synthesized DNA strands at a replication fork. The lagging strand is synthesized discontinuously and therefore its completion lags behind the second, or leading, strand.

**lambda bacteriophage**    A viral parasite that infects bacteria; widely used as a DNA cloning vector.

**leading strand**    One of the two newly synthesized DNA strands at a replication fork. The leading strand is made by continuous synthesis in the 5' to 3' direction.

**leucine zipper**    A structural motif of DNA binding proteins, in which two identical proteins are joined together at regularly spaced leucine residues, much like a zipper, to form a dimer.

**leukemia**    Cancer of white blood cells.

**lipid bilayer**    Two closely aligned sheets of phospholipids that form the core structure of all cell membranes. The two layers are aligned such that the hydrophobic tails are interior, while the hydrophilic head groups are exterior on both surfaces.

**liposome**    An artificial lipid bilayer vesicle used in membrane studies and as an artificial gene therapy vector.

**locus**    A term from genetics that refers to the position of a gene along a chromosome. Different alleles of the same gene occupy the same locus.

**long-term potentiation (LTP)**    A physical remodeling of synaptic junctions that receive continuous stimulation.

**lumen**    A cavity completely surrounded by epithelial cells.

**lymphocyte**    A type of white blood cell that is involved in the adaptive immune response. There are two kinds of lymphocytes: T lymphocytes and B lymphocytes. T lymphocytes (T cells) mature in the thymus and attack invading microbes directly. B lymphocytes (B cells) mature in the bone marrow and make antibodies that are designed to immobilize or destroy specific microbes or antigens.

**lysis**    The rupture of the cell membrane followed by death of the cell.

**lysosome**    Membrane-bounded organelle of eukaryotes that contains powerful digestive enzymes.

**macromolecule**    A very large molecule that is built from smaller molecular subunits. Common examples are DNA, proteins, and polysaccharides.

**magnetic resonance imaging (MRI)**   A procedure in which radio waves and a powerful magnet linked to a computer are used to create detailed pictures of areas inside the body. These pictures can show the difference between normal and diseased tissue. MRI makes better images of organs and soft tissue than other scanning techniques, such as CT or X-ray. MRI is especially useful for imaging the brain, spine, the soft tissue of joints, and the inside of bones. Also called nuclear magnetic resonance imaging.

**major histocompatibility complex**   Vertebrate genes that code for a large family of cell-surface glycoproteins that bind foreign antigens and present them to T cells to induce an immune response.

**malignant**   Refers to the functional status of a cancer cell that grows aggressively and is able to metastasize, or colonize, other areas of the body.

**mammography**   The use of X-rays to create a picture of the breast.

**MAP-kinase (mitogen-activated protein kinase)**   A protein kinase that is part of a cell proliferation–inducing signaling pathway.

**M-cyclin**   A eukaryote enzyme that regulates mitosis.

**meiosis**   A special form of cell division by which haploid gametes are produced. This is accomplished with two rounds of cell division, but only one round of DNA replication.

**melanocyte**   A skin cell that produces the pigment melanin.

**membrane**   The lipid bilayer and the associated glycocalyx that surround and enclose all cells.

**membrane channel**   A protein complex that forms a pore or channel through the membrane for the free passage of ions and small molecules.

**membrane potential**   A buildup of charged ions on one side of the cell membrane establishes an electrochemical gradient that is measured in millivolts (mV); an important characteristic of neurons as it provides the electrical current, when ion channels open, that enable these cells to communicate with one another.

**mesoderm**   An embryonic germ layer that gives rise to muscle, connective tissue, bones, and many internal organs.

**messenger RNA (mRNA)**   An RNA transcribed from a gene that is used as the gene template by the ribosomes and other components of the translation machinery to synthesize a protein.

**metabolism** The sum total of the chemical processes that occur in living cells.

**metaphase** The stage of mitosis at which the chromosomes are attached to the spindle but have not begun to move apart.

**metaphase plate** Refers to the imaginary plane established by the chromosomes as they line up at right angles to the spindle poles.

**metaplasia** A change in the pattern of cellular behavior that often precedes the development of cancer.

**metastasis** Spread of cancer cells from the site of the original tumor to other parts of the body.

**meter** Basic unit in the metric system; equal to 39.4 inches or 1.09 yards.

**methyl group (-CH$_3$)** Hydrophobic chemical group derived from methane; occurs at the end of a fatty acid.

**micrograph** Photograph taken through a light, or electron, microscope.

**micrometer (µm or micron)** Equal to $10^{-6}$ meters.

**microtubule** A fine cylindrical tube made of the protein tubulin, forming a major component of the eukaryote cytoskeleton.

**millimeter (mm)** Equal to $10^{-3}$ meters.

**mitochondrion (plural mitochondria)** Eukaryote organelle, formerly free living, that produces most of the cell's ATP.

**mitogen** A hormone or signaling molecule that stimulates cells to grow and divide.

**mitosis** Division of a eukaryotic nucleus; from the Greek *mitos,* meaning a thread, in reference to the threadlike appearance of interphase chromosomes.

**mitotic chromosome** Highly condensed duplicated chromosomes held together by the centromere. Each member of the pair is referred to as a sister chromatid.

**mitotic spindle** Array of microtubules, fanning out from the polar centrioles, and connecting to each of the chromosomes.

**molecule** Two or more atoms linked together by covalent bonds.

**monoclonal antibody** An antibody produced from a B cell–derived clonal line. Since all of the cells are clones of the original B cell, the antibodies produced are identical.

**monocyte** A type of white blood cell that is involved in the immune response.

**motif**    An element of structure or pattern that may be a recurring domain in a variety of proteins.

**M phase**    The period of the cell cycle (mitosis or meiosis) when the chromosomes separate and migrate to the opposite poles of the spindle.

**multipass transmembrane protein**    A membrane protein that passes back and forth across the lipid bilayer.

**multipotency**    The property by which an undifferentiated animal cell can give rise to many of the body's cell types.

**mutant**    A genetic variation within a population.

**mutation**    A heritable change in the nucleotide sequence of a chromosome.

**myelin sheath**    Insulation applied to the axons of neurons. The sheath is produced by oligodendrocytes in the central nervous system and by Schwann cells in the peripheral nervous system.

**myeloid cell**    White blood cells other than lymphocytes.

**myoblast**    Muscle precursor cell; many myoblasts fuse into a syncytium, containing many nuclei, to form a single muscle cell.

**myocyte**    A muscle cell.

**NAD (nicotine adenine dinucleotide)**    Accepts a hydride ion ($H^-$), produced by the Krebs cycle, forming NADH, the main carrier of electrons for oxidative phosphorylation.

**NADH dehydrogenase**    Removes electrons from NADH and passes them down the electron transport chain.

**nanometer (nm)**    Equal to $10^{-9}$ meters or $10^{-3}$ microns.

**National Institutes of Health (NIH)**    A biomedical research center that is part of the U.S. Department of Health and Human Services. NIH consists of more than 25 research institutes, including the National Institute of Aging (NIA) and the National Cancer Institute (NCI). All of the institutes are funded by the federal government.

**natural killer cell (NK cell)**    A lymphocyte that kills virus-infected cells in the body; also kills foreign cells associated with a tissue or organ transplant.

**neuromodulator**    A chemical released by neurons at a synapse that modifies the behavior of the targeted neuron(s).

**neuromuscular junction**    A special form of synapse between a motor neuron and a skeletal muscle cell.

**neuron**  A cell specially adapted for communication that forms the nervous system of all animals.

**neurotransmitter**  A chemical released by the synapse that activates the targeted neuron.

**non–small cell lung cancer**  A group of lung cancers that includes squamous cell carcinoma, adenocarcinoma, and large cell carcinoma. The small cells are endocrine cells.

**northern blotting**  A technique for the study of gene expression. Messenger RNA (mRNA) is fractionated on an agarose gel and then transferred to a piece of nylon filter paper (or membrane). A specific mRNA is detected by hybridization with a labeled DNA or RNA probe. The original blotting technique invented by E. M. Southern inspired the name. Also known as RNA blotting.

**nuclear envelope**  The double membrane (two lipid bilayers) enclosing the cell nucleus.

**nuclear localization signal (NLS)**  A short amino acid sequence located on proteins that are destined for the cell nucleus, after they are translated in the cytoplasm.

**nuclei acid**  DNA or RNA, a macromolecule consisting of a chain of nucleotides.

**nucleolar organizer**  Region of a chromosome containing a cluster of ribosomal RNA genes that gives rise to the nucleolus.

**nucleolus**  A structure in the nucleus where ribosomal RNA is transcribed and ribosomal subunits are assembled.

**nucleoside**  A purine or pyrimidine linked to a ribose or deoxyribose sugar.

**nucleosome**  A beadlike structure, consisting of histone proteins.

**nucleotide**  A nucleoside containing one or more phosphate groups linked to the 5' carbon of the ribose sugar. DNA and RNA are nucleotide polymers.

**nucleus**  Eukaryote cell organelle that contains the DNA genome on one or more chromosomes.

**oligodendrocyte**  A myelinating glia cell of the vertebrate central nervous system.

**oligo labeling**  A method for incorporating labeled nucleotides into a short piece of DNA or RNA. Also known as the random-primer labeling method.

**oligomer** A short polymer, usually consisting of amino acids (oligopeptides), sugars (oligosaccharides), or nucleotides (oligonucleotides); taken from the Greek word *oligos,* meaning few or little.

**oncogene** A mutant form of a normal cellular gene, known as a proto-oncogene, that can transform a cell to a cancerous phenotype.

**oocyte** A female gamete or egg cell.

**operator** A region of a prokaryote chromosome that controls the expression of adjacent genes.

**operon** Two or more prokaryote genes that are transcribed into a single mRNA.

**organelle** A membrane-bounded structure, occurring in eukaryote cells, that has a specialized function. Examples are the nucleus, Golgi complex, and endoplasmic reticulum.

**osmosis** The movement of solvent across a semipermeable membrane that separates a solution with a high concentration of solutes from one with a low concentration of solutes. The membrane must be permeable to the solvent but not to the solutes. In the context of cellular osmosis, the solvent is always water, the solutes are ions and molecules, and the membrane is the cell membrane.

**osteoblast** Cells that form bones.

**ovulation** Rupture of a mature follicle with subsequent release of a mature oocyte from the ovary.

**oxidative phosphorylation** Generation of high-energy electrons from food molecules that are used to power the synthesis of ATP from ADP and inorganic phosphate. The electrons are eventually transferred to oxygen, to complete the process; occurs in bacteria and mitochondria.

***p53*** A tumor suppressor gene that is mutated in about half of all human cancers. The normal function of the *p53* protein is to block passage through the cell cycle when DNA damage is detected.

**parthenogenesis** A natural form of animal cloning whereby an individual is produced without the formation of haploid gametes and the fertilization of an egg.

**pathogen** An organism that causes disease.

**PCR (polymerase chain reaction)** A method for amplifying specific regions of DNA by temperature cycling a reaction mixture containing the template, a heat-stable DNA polymerase, and replication primers.

**peptide bond**   The chemical bond that links amino acids together to form a protein.

**pH**   Measures the acidity of a solution as a negative logarithmic function (p) of $H^+$ concentration (H). Thus, a pH of 2.0 ($10^{-2}$ molar $H^+$) is acidic, whereas a pH of 8.0 ($10^{-8}$ molar $H^+$) is basic.

**phagocyte**   A cell that engulfs other cells or debris by phagocytosis.

**phagocytosis**   A process whereby cells engulf other cells or organic material by endocytosis. A common practice among protozoans and cells of the vertebrate immune system; from the Greek *phagein*, "to eat."

**phenotype**   Physical characteristics of a cell or organism.

**phosphokinase**   An enzyme that adds phosphate to proteins.

**phospholipid**   The kind of lipid molecule used to construct cell membranes. Composed of a hydrophilic head-group, phosphate, glycerol, and two hydrophobic fatty acid tails.

**phosphorylation**   A chemical reaction in which a phosphate is covalently bonded to another molecule.

**photoreceptor**   A molecule or cell that responds to light.

**photosynthesis**   A biochemical process in which plants, algae, and certain bacteria use energy obtained from sunlight to synthesize macromolecules from $CO_2$ and $H_2O$.

**phylogeny**   The evolutionary history of a group of organisms, usually represented diagrammatically as a phylogenetic tree.

**pinocytosis**   A form of endocytosis whereby fluid is brought into the cell from the environment.

**pixel**   One element in a data array that represents an image or photograph.

**placebo**   An inactive substance that looks the same and is administered in the same way as a drug in a clinical trial.

**plasmid**   A minichromosome, often carrying antibiotic-resistant genes, that occurs naturally among prokaryotes; used extensively as a DNA cloning vector.

**platelet**   A cell fragment derived from megakaryocytes and lacking a nucleus that is present in the bloodstream and is involved in blood coagulation.

**ploidy**   The total number of chromosomes (n) that a cell has. Ploidy is also measured as the amount of DNA (C) in a given cell, relative to a haploid nucleus of the same organism. Most organisms are diploid, having two sets of chromosomes, one from each parent, but there is great variation among plants and animals. The silk gland of the moth *Bombyx mori,* for example, has cells that are extremely polyploid, reaching values of 100,000C, flowers are often highly polyploid, and vertebrate hepatocytes may be 16C.

**pluripotency**　The property by which an undifferentiated animal cell can give rise to most of the body's cell types.

**poikilotherm**　An animal incapable of regulating its body temperature independent of the external environment. It is for this reason that such animals are restricted to warm tropical climates.

**point mutation**　A change in DNA, particularly in a region containing a gene, that alters a single nucleotide.

**polarization**　A term used to describe the reestablishment of a sodium ion gradient across the membrane of a neuron. Polarization followed by depolarization is the fundamental mechanism by which neurons communicate with one another.

**polyacrylamide**　A tough polymer gel that is used to fractionate DNA and protein samples.

**polyploid**　Possessing more than two sets of homologous chromosomes.

**polyploidization**　DNA replication in the absence of cell division; provides many copies of particular genes and thus occurs in cells that highly active metabolically (see ploidy).

**portal system**　A system of liver vessels that carries liver enzymes directly to the digestive tract.

**post-mitotic**　Refers to a cell that has lost the ability to divide.

**probe**　Usually a fragment of a cloned DNA molecule that is labeled with a radioisotope or fluorescent dye, and used to detect specific DNA or RNA molecules on southern or northern blots.

**progenitor cell**　A cell that has developed from a stem cell but can still give rise to a limited variety of cell types.

**proliferation**　A process whereby cells grow and divide.

**promoter**　A DNA sequence to which RNA polymerase binds to initiate gene transcription.

**prophase**　The first stage of mitosis; the chromosomes are duplicated and are beginning to condense but are attached to the spindle.

**protein**　A major constituent of cells and organisms. Proteins, made by linking amino acids together, are used for structural purposes and regulate many biochemical reactions in their alternative role as enzymes. Proteins range in size from just a few amino acids to more than 200.

**protein glycosylation**     The addition of sugar molecules to a protein.

**proto-oncogene**     A normal gene that can be converted to a cancer-causing gene (oncogene) by a point mutation or through inappropriate expression.

**protozoa**     Free-living, single-cell eukaryotes that feed on bacteria and other microorganisms. Common examples are *Paramecium* and *Amoeba*. Parasitic forms inhabit the digestive and urogenital tract of many animals, including humans.

**P-site**     The binding site on the ribosome for the growing protein (or peptide) chain.

**purine**     A nitrogen-containing compound that is found in RNA and DNA. Two examples are adenine and guanine.

**pyrimidine**     A nitrogen-containing compound found in RNA and DNA. Examples are cytosine, thymine, and uracil (RNA only).

**radioactive isotope**     An atom with an unstable nucleus that emits radiation as it decays.

**randomized clinical trial**     A study in which the participants are assigned by chance to separate groups that compare different treatments; neither the researchers nor the participants can choose which group. Using chance to assign people to groups means that the groups will be similar and that the treatments they receive can be compared objectively. At the time of the trial, it is not known which treatment is best.

**random primer labeling**     A method for incorporating labeled nucleotides into a short piece of DNA or RNA.

**reagent**     A chemical solution designed for a specific biochemical or histochemical procedure.

**recombinant DNA**     A DNA molecule that has been formed by joining two or more fragments from different sources.

**refractive index**     A measure of the ability of a substance to bend a beam of light expressed in reference to air that has, by definition, a refractive index of 1.0.

**regulatory sequence**     A DNA sequence to which proteins bind that regulate the assembly of the transcriptional machinery.

**replication bubble**   Local dissociation of the DNA double helix in preparation for replication. Each bubble contains two replication forks.

**replication fork**   The Y-shape region of a replicating chromosome; associated with replication bubbles.

**replication origin (origin of replication, ORI)**   The location at which DNA replication begins.

**respiratory chain (electron transport chain)**   A collection of iron- and copper-containing proteins, located in the inner mitochondrion membrane, that use the energy of electrons traveling down the chain to synthesize ATP.

**restriction enzyme**   An enzyme that cuts DNA at specific sites.

**restriction map**   The size and number of DNA fragments obtained after digesting with one or more restriction enzymes.

**retrovirus**   A virus that converts its RNA genome to DNA once it has infected a cell.

**reverse transcriptase**   An RNA-dependent DNA polymerase. This enzyme synthesizes DNA by using RNA as a template, the reverse of the usual flow of genetic information from DNA to RNA.

**ribosomal RNA (rRNA)**   RNA that is part of the ribosome and serves both a structural and functional role, possibly by catalyzing some of the steps involved in protein synthesis.

**ribosome**   A complex of protein and RNA that catalyzes the synthesis of proteins.

**rough endoplasmic reticulum (rough ER)**   Endoplasmic reticulum that has ribosomes bound to its outer surface.

*Saccharomyces*   Genus of budding yeast that are frequently used in the study of eukaryote cell biology.

**sarcoma**   Cancer of connective tissue.

**Schwann cell**   Glia cell that produces myelin in the peripheral nervous system.

**screening**   Checking for disease when there are no symptoms.

**senescence**   Physical and biochemical changes that occur in cells and organisms with age; from the Latin word *senex,* meaning "old man" or "old age."

**signal transduction**   A process by which a signal is relayed to the interior of a cell where it elicits a response at the cytoplasmic or nuclear level.

**smooth muscle cell**   Muscles lining the intestinal tract and arteries; lack the striations typical of cardiac and skeletal muscle, giving a smooth appearance when viewed under a microscope.

**somatic cell**   Any cell in a plant or animal except those that produce gametes (germ cells or germ cell precursors).

**somatic cell nuclear transfer**   Animal cloning technique whereby a somatic cell nucleus is transferred to an enucleated oocyte. Synonymous with cell nuclear transfer or replacement.

**southern transfer**   The transfer of DNA fragments from an agarose gel to a piece of nylon filter paper. Specific fragments are identified by hybridizing the filter to a labeled probe; invented by the Scottish scientist E. M. Southern, in 1975; also known as DNA blotting.

**stem cell**   Pluripotent progenitor cell found in embryos and various parts of the body that can differentiate into a wide variety of cell types.

**steroid**   A hydrophobic molecule with a characteristic four-ringed structure. Sex hormones, such as estrogen and testosterone, are steroids.

**structural gene**   A gene that codes for a protein or an RNA; distinguished from regions of the DNA that are involved in regulating gene expression but are noncoding.

**synapse**   A neural communication junction between an axon and a dendrite. Signal transmission occurs when neurotransmitters, released into the junction by the axon of one neuron, stimulate receptors on the dendrite of a second neuron.

**syncytium**   A large multinucleated cell. Skeletal muscle cells are syncytiums produced by the fusion of many myoblasts.

**syngeneic transplants**   A patient receives tissue or an organ from an identical twin.

**tamoxifen**   A drug that is used to treat breast cancer. Tamoxifen blocks the effects of the hormone estrogen in the body. It belongs to the family of drugs called antiestrogens.

**T cell (T lymphocyte)**   A white blood cell involved in activating and coordinating the immune response.

**telomere**   The end of a chromosome; replaced by the enzyme telomerase with each round of cell division to prevent shortening of the chromosomes.

**telophase**   The final stage of mitosis in which the chromosomes decondense and the nuclear envelope reforms.

**template**   A single strand of DNA or RNA whose sequence serves as a guide for the synthesis of a complementary, or daughter, strand.

**therapeutic cloning**   The cloning of a human embryo for the purpose of harvesting the inner cell mass (embryonic stem cells).

**topoisomerase**   An enzyme that makes reversible cuts in DNA to relieve strain or to undo knots.

**totipotency**   The property by which an undifferentiated animal cell can give rise to all of the body's cell types. The fertilized egg and blastomeres from an early embryo are the only cells possessing this ability.

**transcription**   The copying of a DNA sequence into RNA, catalyzed by RNA polymerase.

**transcription factor**   A general term referring to a wide assortment of proteins needed to initiate or regulate transcription.

**transfection**   Introduction of a foreign gene into a eukaryote or prokaryote cell.

**transfer RNA (tRNA)**   A collection of small RNA molecules that transfer an amino acid to a growing polypeptide chain on a ribosome. There is a separate tRNA for amino acid.

**transgenic organism**   A plant or animal that has been transfected with a foreign gene.

**trans Golgi network**   The membrane surfaces where glycoproteins and glycolipids exit the Golgi complex in transport vesicles.

**translation**   A ribosome-catalyzed process whereby the nucleotide sequence of a mRNA is used as a template to direct the synthesis of a protein.

**transposable element (transposon)**   A segment of DNA that can move from one region of a genome to another.

**ultrasound (ultrasonography)**   A procedure in which high-energy sound waves (ultrasound) are bounced off internal tissues or organs producing echoes that are used to form a picture of body tissues (a sonogram).

**umbilical cord blood stem cells**   Stem cells, produced by a human fetus and the placenta, that are found in the blood that passes from the placenta to the fetus.

**vector**   A virus or plasmid used to carry a DNA fragment into a bacterial cell (for cloning) or into a eukaryote to produce a transgenic organism.

**vesicle**    A membrane-bounded bubble found in eukaryote cells. Vesicles carry material from the ER to the Golgi and from the Golgi to the cell membrane.

**virus**    A particle containing an RNA or DNA genome surrounded by a protein coat. Viruses are cellular parasites that cause many diseases.

**western blotting**    The transfer of protein from a polyacrylamide gel to a piece of nylon filter paper. Specific proteins are detected with labeled antibodies. The name was inspired by the original blotting technique invented by the Scottish scientist E. M. Southern in 1975; also known as protein blotting.

**xenogeneic transplants (xenograft)**    A patient receives tissue or an organ from an animal of a different species.

**yeast**    Common term for unicellular eukaryotes that are used to brew beer and make bread. *Saccharomyces cerevisiae* (baker's yeast) are also widely used in studies on cell biology.

**zygote**    A diploid cell produced by the fusion of a sperm and egg.

 # Further Resources

## BOOKS

Alberts, Bruce, et al. *Essential Cell Biology.* 3rd ed. New York: Garland Publishing, 2009. A basic introduction to cellular structure and function that is suitable for high school students.

Alberts, Bruce, Alexander Johnson, Julian Lewis, Martin Raff, Keith Roberts, and Peter Walter. *Molecular Biology of the Cell.* 5th ed. New York: Garland Publishing, 2008. Advanced coverage of cell biology that is suitable for senior high school and university students.

Brooks, George, Karen Carroll, Janet Butel, and Stephen Morse. *Medical Microbiology.* 24th ed. New York: McGraw-Hill, 2007. This book devotes an entire section to virology, which includes general properties of viruses, taxonomy, and viral diseases.

Ganong, William. *"Review of Medical Physiology."* 22nd ed. New York: McGraw-Hill, 2005. A well-written overview of human physiology, beginning with the basic properties of cells and tissues.

Krause, W. J. *Krause's Essential Human Histology for Medical Students.* Boca Raton, Fla.: Universal Publishers, 2005. This book goes well with histology videos provided free on Google video.

Oldstone, Michael. *Viruses, Plagues, and History.* New York: Oxford University Press, 2010. A comprehensive look at the history of viral diseases and modern attempts to control the epidemics and pandemics.

Panno, Joseph. *Aging: Modern Theories and Therapies.* Rev. ed. New York: Facts On File, 2010. Explains why and how people age, and how gene therapy may be used to reverse or modify the process.

―――. *Animal Cloning: The Science of Nuclear Transfer.* Rev. ed. New York: Facts On File, 2010. Medical applications of cloning technology are discussed including therapeutic cloning.

―――. *Cancer: The Role of Genes, Lifestyle, and Environment.* Rev. ed. New York: Facts On File, 2010. The basic nature of cancer written for the general public and young students.

―――. *The Cell: Exploring Nature's First Life-form.* Rev. ed. New York: Facts On File, 2010. Everything you need to know about the cell without having to read a 1,000-page textbook.

―――. *Gene Therapy: Treatments and Cures for Genetic Diseases.* Rev. ed. New York: Facts On File, 2010. Discusses not only the great potential of this therapy, but also its dangers and its many failures.

―――. *The Immune System: Nature's Way of Fighting Disease.* New York: Facts On File, forthcoming 2012. A detailed overview of the human immune system with special emphasis on its ability to deal with viral infections.

―――. *Stem Cell Research: Medical Applications & Ethical Controversies.* Rev. ed. New York: Facts On File, 2010. All about a special type of cell, the stem cell, and its use in medical therapies.

## JOURNALS AND MAGAZINES

Bainbridge, James, et al. "Effect of Gene Therapy on Visual Function in Leber's Congenital Amaurosis." *New England Journal of Medicine* 358 (May 22, 2008): 2,231–2,239. A British group, led by Dr. Robin Ali, used gene therapy to restore the night vision in a patient suffering from retinal dystrophy.

Cavazzano-Calvo, Marina, and Alain Fischer. "Gene Therapy for Severe Combined Immunodeficiency: Are We There Yet?"

*Journal of Clinical Investigation* 117 (June 2007): 1,456–1,465. An interesting review article that summarizes the successes and failures of SCID-X1 gene therapy.

Church, George. "Genomes for All." *Scientific American* 294 (2006): 46–54. This article discusses fast and cheap DNA sequencers that could make it possible for everyone to have their genome sequenced, giving new meaning to personalized medicine.

Danovaro, Roberto, et al. "Major Viral Impact on the Functioning of Benthic Deep-Sea Ecosystems." *Nature* 454 (August 2008): 1,084–1,087. This paper discusses the extent to which viruses control the ecology and especially the bacterial populations of deep-sea beds.

Fretag, Svend, et al. "Phase I Trial of Replication-competent Adenovirus-mediated Suicide Gene Therapy Combined with IMRT for Prostate Cancer." *Molecular Therapy* 15 (May 2007): 1,016–1,023. Gene therapy is used to induce the destruction of prostate cancer cells.

Garcon, Nathalie, and Michel Goldman. "Boosting Vaccine Power." *Scientific American* (October 2009): 72–79. With HIV still resisting current vaccines, researchers are trying to find ways to boost the effectiveness of a vaccine by making the immune system more responsive to their presence.

Al-Hashimi, Hashim. "Aerial View of the HIV Genome." *Nature* 460 (August 2009): 696–698. This is a news article that profiles research into the detailed structure of the HIV genome.

Kaplitt, Michael, et al. "Safety and Tolerability of Gene Therapy with an Adeno-associated Virus (AAV) Borne *GAD* Gene for Parkinson's Disease: An Open Label, Phase I Trial." *Lancet* 369 (June 23, 2007): 2,097–2,105. The first time gene therapy has been used to boost the production of a neurotransmitter in order to alleviate some of the symptoms of PD.

Morgan, Richard, et al. "Cancer Regression in Patients after Transfer of Genetically Engineered Lymphocytes." *Science* 314

(October 6, 2006): 126–129. Steven Rosenberg's team at NCI treat melanoma with gene therapy.

Nettelbeck, Dirk, and David Curiel. "Tumor-Busting." *Scientific American* (October 2003): 68–75. This article discusses the use of adenoviruses (virotherapy) to treat cancer.

Rohwer, Forest, and Rebecca Thurber. "Viruses Manipulate the Marine Environment." *Nature* 459 (May 2009): 207–212. A fascinating look at the many ways in which viruses influence the behavior and evolution of creatures in the sea.

Soares, Christine. "Pandemic Payoff." *Scientific American* (November 2009): 19–20. The author discusses the possibility that the virulence of the Spanish flu of 1918 may be the reason why the swine flu of 2009 was so tame.

Stevenson, Mario. "Can HIV Be Cured?" *Scientific American* (November 2008): 78–83. An interesting discussion of several new strategies that might lead to an effective anti-HIV therapy.

Suttle, Curtis. "Viruses in the Sea." *Nature* 437 (September 2005): 356–361. A comprehensive article that discusses the central role that viruses play in the ecology of the sea.

Varala, Mariana. "Friendly Viruses." *Annals of the New York Academy of Sciences* 1,178 (October 2009): 157–172. This paper discusses some of the benefits that humans and other animals derive from retroviruses.

Villarreal, Luis. "Are Viruses Alive?" *Scientific American* 291 (2004): 101–105. A fascinating discussion of what it means to be a virus.

Virgin, Herbert, and Bruce Walker. "Immunology and the Elusive AIDS Vaccine." *Nature* 464 (March 11, 2010): 224–231. The authors discuss the problems associated with the development of an effective AIDS vaccine.

Wolfe, Nathan. "Preventing the Next Pandemic." *Scientific American* (April 2009): 76–81. Wild animals are the source of many human viral diseases. By tracking the course of these diseases in the wild it may be possible to prevent them from becoming human pandemics.

## ARTICLES ON THE INTERNET

Clarke, Toni. "Genzyme bets on gene therapy as others steer clear." *Reuters,* (6/21/07). Available online. URL: http://www.reuters. com/article/health-SP/idUSN2135767920070621. Accessed May 15, 2010. The biotech company, Genzyme, is developing a gene therapy for Parkinson's disease.

Human Genome Project. "Gene Therapy." Available online. URL: http://www.ornl.gov/sci/techresources/Human_Genome/ medicine/genetherapy.shtml. Accessed May 15, 2010. This article discusses gene therapy and recent developments in the field.

Murphy, Clare. "Why Don't We Vaccinate Against Chicken pox?" Available online. URL: http://news.bbc.co.uk/2/hi/ health/8557236.stm. Accessed May 15, 2010. This article from the BBC discusses some of the reasons why the British refuse to use the chicken pox vaccine.

National Institutes of Health. "Stem Cell Information." Available online. URL: http://stemcells.nih.gov/index.asp. Accessed May 15, 2010. Covers both the scientific and political aspects of stem cell research.

Nature.com. "Swine Flu." Available online. URL: http://www. nature.com/news/specials/swineflu/index.html. Accessed May 15, 2010. A collection of articles dealing with the swine flu pandemic of 2009.

Wong, Edward. "China's Tough Flu Measures Appear to Be Effective." *New York Times* (11/12/09). Available online. URL: http:// www.nytimes.com/2009/11/12/world/asia/12chinaflu.html?_ r=1&scp=1&sq=China's%20Tough%20Flu%20Measures%20 Appear%20to%20Be%20Effective&st=cse. Accessed May 15, 2010. China introduced some tough measures to control the swine flu epidemic.

Wren, Jonathan, et al. "Plant Virus Biodiversity and Ecology." *PLOS Biology* 4 (2006): 314–315. Available online. URL: http://biology.plosjournals.org/perlserv/?request=get-document&doi=10.1371/journal. pbio.0040080. Accessed May 15, 2010. This article discusses the number and types of viruses that infect terrestrial organisms.

## WEB SITES

Centers for Disease Control and Prevention. Available online. URL: http://www.cdc.gov/. Accessed May 15, 2010. This site provides detailed descriptions of various viral pathogens, such as smallpox, influenza, and HIV.

Department of Energy Human Genome Project. Available online. URL: http://genomics.energy.gov. Accessed May 15, 2010. Covers every aspect of the human genome project and current genome programs with extensive color illustrations.

Genetic Science Learning Center at the University of Utah. Available online. URL: http://learn.genetics.utah.edu/. Accessed May 15, 2010. An excellent resource for beginning students. This site contains information and illustrations covering basic cell biology, gene therapy, animal cloning, stem cells, and other new biology topics.

Google Video. Available online. URL: http://video.google.com/videosearch?q=viruses+in+the+body&www_google_domain=www.google.com&tbo=p&tbs=vid%3A1&source=vgc&aq=0&oq=viruses+in#. Accessed May 15, 2010. This site contains many videos covering viruses and viral infections.

National Cancer Institute. Available online. URL: http://www.cancer.gov/. Accessed May 15, 2010. This site, established by the National Institutes of Health, covers basic cancer information and links to gene therapy clinical trials.

National Center for Biotechnology Information (NCBI). Available online. URL: http://www.ncbi.nlm.nih.gov. Accessed May 15,

2010. This is an excellent resource for anyone interested in biology. The NCBI provides access to GenBank (DNA sequences), literature databases (Medline and others), molecular databases, and topics dealing with genomic biology. With the literature database, for example, anyone can access Medline's 11,000,000 biomedical journal citations to research biomedical questions. Many of these links provide free access to full-length research papers.

National Institute of Health. Available online. URL: http://www.nih.gov. Accessed May 15, 2010. The NIH posts information on their Web site that covers a broad range of topics including general health information, virology, stem cell biology, aging, and much more.

National Network for Immunization Information. Available online. URL: http://www.immunizationinfo.org/about-nnii. Accessed May 15, 2010. This site provides up-to-date information regarding immunization policies in the United States.

Nobel Foundation. Available online. URL: http://nobelprize.org/nobelfoundation/index.html. Accessed May 15, 2010. This site provides biographies and photos of Nobel Prize winners.

UNAIDS. Available online. URL: http://www.unaids.org/en/KnowledgeCentre/HIVData/EpiUpdate/EpiUpdArchive/2009/default .asp. Accessed May 15, 2010. This site, established by the United Nations, provides detailed annual updates regarding the AIDS epidemic.

United States Food and Drug Administration. Available online. URL: http://www.fda.gov/BiologicsBloodVaccines/default.htm. Accessed May 15, 2010. Provides extensive coverage of general health issues and regulations, including updates regarding swine flu and the production of the seasonal flu vaccine for 2010 and 2011.

Virology Journal. Available online. URL: http://www.virology.net/garryfavweb.html. Accessed May 15, 2010. This site provides extensive coverage of basic virology, with many pictures and links to additional resources.

World Health Organization (WHO). Available online. URL: http://www.who.int/en. Accessed May 15, 2010. This site provides detailed information regarding efforts to eliminate infectious diseases around the world.

# Index